PITCH, SKETCH, LAUNCH

"Krewson has a way of making complex ideas easily understandable, and *Pitch, Sketch, Launch* is no different. In it, he offers pragmatic, common-sense insights that have the capacity to generate fast—and tangible—results."

—**MARK BROOKS**, CIO, Reinsurance Group of America

"For as creative as software and product development is, frequently the industry gets stuck in 'following patterns' and avoiding uncertainty. Krewson, in an engaging and often hilarious format, shows us better ways—ways that increase innovation, foster deep collaboration, and just make work more enjoyable. Looking for ways to drive more creativity? You found them."

—**JOEL TOSI**, Cofounder, Dojo & Co.

"This book is for the person who can't watch *The Office* because it's too much like a documentary. By dismantling the 'we have always done it this way' mantra, Krewson walks you through insightful ways to rethink how to get your product to 'done' efficiently, creatively, and with reenergized excitement in your team. Fair warning: Page after page of nodding in affirmation may result in a stiff neck."

—**JOE DENT**, Software Product Leader

"*Pitch, Sketch, Launch* is a refreshing take on productivity, creativity, teamwork, and innovation. Krewson makes a compelling case for shifting decision-making from top-down directives to team-driven solutions, fostering agility and ownership. It challenges the traditional obsession with 'busyness' and instead highlights how unnecessary layers of oversight actually hinder true productivity. Most importantly, it reframes unpredictability—not as a problem to be solved, but as a creative force to be embraced. A must-read for anyone looking to unlock better ways of leading and working."

—**DAVID SHARP**, VP, Business Technology Enablement

"In *Pitch, Sketch, Launch*, Krewson and Kutner have achieved something remarkable—a fresh, practical approach to innovation. Their work draws parallels between sketch comedy production and software development, showing how the nimble, collaborative methods of shows like *Saturday Night Live* can transform rigid corporate structures into dynamic idea factories. Whether you're building software or rethinking an organizational process, this book will change how you approach teamwork, ideation, and execution. A must-read for anyone who wants their business to stay relevant in a rapidly evolving marketplace."

—**ANGIE ECHELE**, CMO

"Krewson delivers a fresh perspective on how product teams deliver in a way that shares his deep experience yet is profoundly engaging and approachable. By drawing parallels between software development and sketch comedy, he reframes how teams work, think, and engage with a completely new—and sometimes surprising—perspective. John's approach is exactly what teams need to be more innovative and collaborative and reach new levels of excellence."

—**JULIA PITLYK**, Director, Marketing Technology, Nestle Health Science

"Krewson skillfully navigates the complexity and traps of product management and value creation. He challenges conventional thinking in favor of creativity, teamwork, and true innovation, and it's mind-opening, motivational, and overall enjoyable to read. Like your favorite film, it's a book you'll want to read over and over again and be eager to recommend and discuss with others."

—**KATE HORAN**, Director of Digital Products and Platforms, Eastman

"John and Rob provide fresh insights into software development by brilliantly connecting the creative process of sketch comedy with agile principles, showing how improvisation, collaboration, and rapid iteration can lead to better, more creative outcomes of software-driven product development."

—**FRANK THOENNES**, CTO, Listrak Inc.

"I've always been struck by how John blends his onstage background with his decades of software product leadership to bring fresh energy to building better products. He shows how sketch comedy uses a 'pitch, sketch, launch' approach to uncover the real 'why' behind product development. It's an inventive and genuinely enjoyable way to reimagine how teams work—one that actually sticks. If you want a more human, creative way to build better products (and have a little fun in the process), this is the read for you."

—**JASON WRUBEL**, Cofounder, Reboot Ventures & Technology Leadership Advisor

"*Pitch, Sketch, Launch* opens a refreshing and engaging perspective on product development with interesting stories that elucidate, educate, and entertain, from both the world of entertainment and the world of business. Anyone involved in product development, whether from a technical, business or operational position, will find abundant value in reading this book.

"I've known and worked with John for over 15 years. His creative approach to problems and products repeatedly yields great results. As any reader of this book will recognize, John delivers a valuable insight into product development, and then does it again, and again, and again. His unique background of sketch comedy and product development provides the kind of learnings you can't find elsewhere."

—**ROBERT LADD**, EVP and CMO, Constellation SaaS

"In a time where a fresh but practical perspective is needed, this book provides just that. With an engaging style and loaded with real-world experience, Krewson challenges both the traditional thinking and 'fluff' and outlines a new pattern for success."

—**CINDY HEMBROCK**, VP, Product Management, Mastercard

"This book is not just insightful—it's a breath of fresh air for anyone navigating creativity, teamwork, and innovation in the business world. Krewson challenges conventional thinking while keeping you engaged and inspired every step of the way."

—**GHAN MEHTA**, VP Strategic Partnerships, Integris Health

"As a tech industry CEO, I know all too well how often development projects end up in a cul-de-sac of despair. I've repeatedly turned to John Krewson for help over the last two decades. He's one of the very few DevOps devotees I've met that actually gets it.

What a fun read! The metaphor and model, based on the comedy process, turns out to be a remarkably insightful take on the challenges of agile development and how to make it work. This is a must read for any frustrated technology leader or senior software developer in 2025."

—**MATTHEW HARRIS**, CEO, US Cloud

PITCH
SKETCH
LAUNCH

What Sketch Comedy Can Teach Us About Product Development

JOHN KREWSON
with ROB KUTNER

WREN HOUSE
press

COPYRIGHT © 2025 JOHN KREWSON
All rights reserved.

PITCH, SKETCH, LAUNCH
What Sketch Comedy Can Teach Us About Product Development
First Edition

ISBN	978-1-967115-02-0	*Hardcover*
	978-1-967115-01-3	*Paperback*
	978-1-967115-00-6	*Ebook*
	978-1-967115-03-7	*Audiobook*

Library of Congress Control Number: 2025907140

To Chrissy, Collin, and Elliot,
my greatest inspirations.

CONTENTS

Foreword *by Connor Ratliff*   xv
Introduction   xxi

ONE   Thinking Funny   1
TWO   Assembling the Troupe   31
THREE   The Pitch Meeting   69
FOUR   Only Listen to the Laugh   101
FIVE   Definitely Ready for Prime Time   123
SIX   The Button   147

Appendix   163
Acknowledgments   185
About the Authors   189

FOREWORD
by CONNOR RATLIFF

The first thing you should know about John Krewson is one of the first things I knew about him: that he came very close to being cast in a leading role (opposite Robert De Niro) in the 1993 motion picture *This Boy's Life*.

Ultimately, the role went to Leonardo DiCaprio, who went on to become one of the biggest movie stars in the world. I often think of a reality in which John would land the role and then automatically go on to star in every movie in Leo's filmography.

John in *Titanic*, screaming, "I'M KING OF THE WORLD!"

John in *The Revenant*, getting mauled by a bear.

John in *Once Upon a Time in Hollywood*, attacking a member of the Manson family in a swimming pool with a flamethrower.

(I also like to imagine that John and Leo basically swapped futures, and Leo did all the stuff John has accomplished, including writing this book.)

The second thing you should know about John is that he created his own company called Sketch Development. They help other companies get better at creating and delivering software. If you need help with that kind of thing, maybe you should think about using them.

The third thing you should know about John is that he asked me to write this foreword. My name is Connor Ratliff—I'm an actor/comedian and creator of the award-winning podcast, *Dead Eyes*. You may have heard my voice on a recent episode of *Bob's Burgers*, or watched me perform long-form improvised comedy at the Upright Citizens Brigade Theatre or various other venues in New York and elsewhere.

I've known John since we were freshmen at the University of Missouri, Columbia—two theater majors on acting scholarships who were both good at acting! John owns a house now, and I recently had a small role on an episode of *Ghosts* on CBS.

During our second semester at MU, John took a computer science class, thinking it would be an easy credit—he had learned to code BASIC when he was ten years old. At that point, John had no sense that this class would more accurately point the way to his career success than his starring roles in MU productions of *Psycho Beach Party* and *"Master Harold"...and the Boys*.

If you carefully peruse vintage SNL clips from the late 1990s, you can find John as a background player. But he caught on very quickly that the showbiz path was a rocky road paved with disaster (can confirm, it is a nightmare!), and that there was a smarter way forward that merged his twin passions of sketch comedy and software development.

You might be asking yourself, "But *why* did John ask *this* guy to write the foreword? What could *he* possibly know about product management?"

In fact, I said words to that effect to John and he immediately began to explain to me the various ways in which I already know a lot about product management. In my career—being a somewhat niche figure in a specific corner of the NY/LA improv comedy scenes with a cult comedy talk show and a critically acclaimed podcast—I have essentially developed a small but devoted audience that is very tuned in to the kinds of projects I do. And when I occasionally make inroads into other, slightly less obscure parts of show business, I have an awareness of how I can hold on to that core following while attracting new people who might also be into it.

I'm not always even aware that *product management* is what I'm doing. It's just a natural part of trying to make a living in a field where my personality and comedic instincts are the core of my profession. In some ways, I *am* the product.

Or part of me is, anyway. And that part—the performative, forward-facing comedy part—has to focus on never letting anything—personal issues, politics, bad habits—get in the way of delivering that product, on stage or on camera. Whether I'm having a good day or a bad day, when it's time to perform, I need to give 100 percent of what is needed, especially since each job is potentially an audition for my *next* job in a highly competitive field. John is trying to bring that same kind of laser focus on product to software, in a world where too many other extraneous factors are coming ahead of it.

Often, I'm part of a team that is working in close collaboration to make a product that is funny and interesting, whether that is performing a fully improvised live show or assembling an episode of a comedy podcast. In those circumstances, we are constantly building on years of training and professional experience—figuring out what works and what doesn't and adapting as necessary. Sometimes that involves failing at comedy, and being unafraid to do so. Allowing yourself the freedom to fail is an important part of the process, because some of the most important discoveries lie just beyond the edge of failure. When people are scared to take risks, they end up with safe, bland, boring, cookie-cutter results—how do you surprise an audience into laughing uproariously if they see everything coming a mile away? You have to be bold if you want to create something that stands out from all the rest.

See? Even though I thought the things in this book had no bearing on what I do in my crazy, ramshackle showbiz career, John immediately got me thinking differently about it, and realizing the various ways that there is overlap between the way our brains approach things to get the best results, regardless of what specific realm we might be working in.

One last thing: Sophomore year at MU, John and I starred in a production of a play called *Marvin's Room*. *Two years later*, they made it into a movie, with Robert De Niro playing my role and Leonardo DiCaprio in John's.

Which means that, in the "alternate reality" where Leo wrote this book, Bobby D. wrote the foreword.

Don't think too hard about the logic behind that—just imagine three-time SNL host Robert De Niro ordering you to read a book about revolutionizing your approach to software development through the insights of the comedy world. "*DO it!*"

Anyway, goodbye. Enjoy the book!

—Connor Ratliff
Actor/Comedian/Friend of John Krewson

INTRODUCTION

It's a brisk October Saturday night in 1976, 11:30 p.m. to be precise. Conditions outside at Rockefeller Center are as calm as can be, but that is in *stark* contrast to the chaos going on upstairs. That's where season two of *Saturday Night Live* is unfolding.

It's Chevy Chase's last episode on the show. Buck Henry is the host, and—as he has the last two times that he's hosted—he's performing in one of the famous John Belushi "Samurai" sketches. This time he's coming to the "Samurai Stockbroker's" office, and complaining to Samurai Belushi about the performance of his stocks. And, they're doing all these gimmicks and gags with—at Belushi's firm insistence—a *real samurai sword*.

The final joke is one where Buck Henry says, "Man, if there was a window in this office, I'd jump through it!" So to create such a window, Belushi takes his sword and starts hacking away at the wall, and he's hitting it so hard with his real sword that, on the back swing, he clips Buck Henry on the forehead, causing the poor guest host to bleed all over the set.

Later that same season, Chevy Chase returned to host an episode, during which he got into a fistfight with Bill Murray (egged on by Belushi) while Billy Joel was performing live on stage.

If you were to pause history and look at what was going on in 1976 in Studio 8H as an objective, rational human being, you would have to conclude: *This is an absolute mess.* People are bleeding. Buck Henry had to run offstage and get stitched up by a doctor during the commercial break. It's just absolute chaos. No one in their right mind would conclude, *Well, this is a show that's going to last for at least fifty years, win over two hundred Emmys, manufacture comedian after comedian, and maintain its spot as the most watched television show in its category for decades.*

Perhaps even more surprising, the way that SNL (and many other sketch shows) do things is teachable, replicable, and directly applicable to your business. This book will show you how, albeit with fewer samurai sword injuries.

But first, as we will kick off each chapter practicing what we preach, a brief sketch of office life:

INT. CONFERENCE ROOM - DAY

EXECUTIVE is holding court over an All-Hands Meeting, peopled by MANAGER and three DEVELOPERS: ROCKSTAR, TEAM PLAYER, and SHY GUY.

> EXECUTIVE
> Alright, folks, I've got some good news and some bad news.

Everyone looks at each other nervously, murmuring.

> EXECUTIVE
> First, the bad news: Our soon-to-be revolutionary product CASTLR—the first-ever app that determines whether your genealogy might entitle *you* to your own castle—is $10 million over budget!

A collective GASP. Manager sits up, smoothing her suit too energetically.

> MANAGER
> Not to worry, Chief. We'll chop off a beta-test round, and make it up on the backend by charging for a premium "Moat" tier.

EXECUTIVE
…and six weeks behind schedule.

Rockstar leaps to his feet.

ROCKSTAR
Then I'm stoked for seven days of all-nighters and turning my blood type to "Red Bull"! WHOO!

Team Player joins him.

TEAM PLAYER
(fist-bumping him) Up here! Sleep is for the weak!

ROCKSTAR
(competitively) Make it *seventeen* days.

EXECUTIVE
Even if we launched tomorrow, our rival OtherCo is this close to shipping "PalaceFindr"!

MANAGER
Ugh! Who's got an idea?

SHY GUY
(pulling out paper with sketches on it) Well, I was just thinking…

> MANAGER
> Is that *guaranteed* to save this product and make us millions?

Intimidated, Shy Guy puts it away.

> SHY GUY
> Um, no…it's stupid…

> EXECUTIVE
> This is DEFCON 9, people! I just got off a call with the client, and he chewed my ear off worse than Tyson! We are *screwed*!

> MANAGER
> What was the…good news?

Executive flings a greasy box onto the table.

> EXECUTIVE
> I brought donuts.

WHAT'S WRONG WITH THIS PICTURE?

Did any of that sound like something that might happen in your office, or your business? It might seem funny to read, but it is anything but a laugh to experience in real life. Moreover, it highlights some of the many problems with the typical way that software is developed these days. And there are many more such dysfunctions standing in the way of high performance. Here's a non-exhaustive (though somewhat exhausting) list:

- It takes forever to get anything built, let alone delivered to a customer.
- Nobody knows how long anything is really going to take to get done, so planning is a nightmare.
- Once it's built, it doesn't work.
- If it does work, customers hate it or don't use it—even though they said they *had* to have it.
- Everyone has their own idea of what the most important things are to work on, and these are all different from each other's.
- When a better idea is discovered, it takes forever to turn the ship around to the new direction.
- Anything that sticks around as part of the product ends up drab, bland, not exciting, mainstream, copycat stuff.
- Everybody hates their job and is just waiting for 5:00 so they can go home for the day.

HOW DID THIS HAPPEN?

How did *software*—one of the twentieth century's, if not humankind's, most innovative creations—stop being creative? When did something that literally comes out of human brains and compels machines to do incredible things our ancestors could never have dreamed of turn into…widget-making?

At the risk of sounding like an ad for a household cleaning product, "There has to be a better way!" And the good news is, there is. But the source of that better way is going to surprise you.

HOW DO WE KNOW?

Okay, you might *not* be surprised to find that one of this book's coauthors, John, is an expert in software development. He's a twenty-five-year veteran of the software industry and the founder and CEO of Sketch Development, helping companies from the Fortune 50 to fifty-person enterprises get better at making and delivering software.

On the other hand, it probably *will* surprise you that he has teamed up with Rob, an expert in TV comedy sketchwriting, with twenty years' experience writing for *The Tonight Show*, *The Daily Show*, and CONAN; and experience teaching graduate-level sketch at Loyola Marymount University's School of Film and Television. John is also a longtime fan and student of sketch comedy and has even performed in it on *Saturday Night Live!*

How did this odd couple come to be? It started when we began to realize that the way comedy sketches are produced professionally could be a helpful model for rethinking how software is made.

We know: That sentence itself might *sound* like the premise for a comedy sketch. But hear us out.

From the outside, sketch comedy may appear casual, frivolous, easily "thrown together," and of little value past the moment. But from a business perspective, sketch comedy is a peer-respectful, sustainable, and *efficacious* way to bring difficult, innovative, short-horizon projects into the world that make everyone happy. Further, sketches are small, independent units of value, which is a smart way to deliver anything.

Does that sound like something you wish your software company was doing?

We suspect it might be. Even so, at this point some of the more Serious Business Types among you might be thinking: *Sure, comedy games are fine for warming up a meeting, or livening up a retreat, but eventually we have to stop laughing and start focusing on getting our product into the end zone.*

And there it is. The sports metaphor, staple of so many business books. While we believe that that beloved rubric is sometimes useful, we also think it's incomplete. Yes, competitive team sports like baseball and football help us see teamwork at its best, and they offer a model of exerting the maximum amount of effort to achieve a goal. But they don't do a good job of accurately capturing the constraints and conditions of a true business environment. They describe a make-believe world full of non-real-world rules monitored by a third-party judge, arbitrarily invented time periods, and externally meaningless metrics.

By contrast, the way sketch comedy is made directly resembles the way software is developed: putting a team into a room with nothing more than a whiteboard and their brains and hoping for a product to come out of it. Or at least the working blueprint of one. Furthermore, sketch comedy (in theory at least) is directly accountable to a feedback system (laughter), which keeps its focus squarely on function, user experience, and market success.

WHAT'S IN IT FOR YOU?
(WHAT THIS BOOK WILL TEACH)

So in this book, we're going to show you how the funny sausage is made, and how you can apply many of those practices to your software business. Or really, any business with occasional blue-sky creative phases.

We'll walk you through the steps of our process we call "Pitch, Sketch, Launch." (Actually, the full series of steps are Pitch, Sketch, Stage, Test, and Launch, but we thought that would be too much to grapple with up top.)

We'll travel through the process by which sketch comedy approaches its product, and provide examples from expert practitioners behind the top shows in the genre. We will show you how your business is likely doing things differently than them, identify the roadblocks you're probably encountering, and show you how a sketch-inspired approach can help you vault over them. (That was technically a track-and-field metaphor, but it's as close to sports as we're going to get.)

We'll also provide examples of how this type of approach is already working in many successful companies, like Pixar and Aflac, putting them ahead of competitors and retaining talented, satisfied, highly collaborative workforces.

In Chapter 1, we'll look at the problematic *mindset* plaguing the culture of many software (and other) companies, holding them back from letting their brilliant people do their best work. We will examine how sketch comedy productions think differently about benchmarks like "efficiency," "productivity," "timeliness," and "succeeding." Our goal will be to help you "free your mind" from unhelpful and time/energy-draining institutional habits that not only divert from budgets and deadlines, but also downgrade the quality of the product.

In Chapter 2, we'll turn our focus to the key ingredient in any corporate concoction: the *people*. What considerations make for better hires, better teams, and better interactions? Spoiler alert—not many of the ones relied upon today. We'll explore how the sketch comedy world views its creator-performers, and some of the unique and impressive ways that their teams work so well together— as well as how they handle it when they don't.

In Chapter 3, we'll take a critical look at the too-often-creativity-dampening *process* by which software gets developed in many contemporary companies. Then we'll dive into the wild ride of the sketch comedy pitch meeting and how that provides a set of tools that can drastically boost output, innovation, and quality control while also adding in the perk of being more fun.

In Chapter 4, we'll steer right into what's likely to be a burning question after all the fun and games: "Okay, but how do I build an actual, viable product around all this?" As with previous chapters, we'll challenge you to reevaluate what's of most value to your process and product, versus your institutional defense systems and policies of habit.

We'll wrap things up in Chapter 5 by demonstrating how *Saturday Night Live* is an effective model for scaling and sustainability. We'll cite how they have built and maintained a fifty-year-franchise with immediate brand recognition over waves of changes and challenges… and how you can apply what they do to what you do.

Many of the concepts in this book are heavily influenced by Agile approaches to software development. But this is not a book about Agile. We have reflected on our (okay, let's be honest, John's) extensive experience applying Agile concepts to real-world environments to determine what really works, and what's just fluff. So you'll recognize some concepts that are Agile at their core, but call out other disciplines, like product management and good old-fashioned project management. The net effect, we hope, is to give you a more fluid set of tools that allows for the building of custom solutions.

Now, before we raise the curtains, some real talk: *We know corporate change is hard.* We also know a lot of what we're about to propose may seem odd or unfamiliar. But we promise the steps suggested in this book will eventually (and sometimes immediately) pay off in terms of a more enjoyable workplace and workflow—and even more down the line in terms of a more nimble company, better products, and higher customer satisfaction.

Why should you believe us? Because over the last ten years, John's company, Sketch Development, has brought this technique to dozens of businesses and achieved great things. They've empowered a large hospital system to revamp its nurse-scheduling software in a way that both the hospital and the nurses love. They've enabled a major pet food company to find millions more dollars of value in its customer data. They've worked alongside a large loyalty system platform to break out of a "vendor-trap" and create software that best suits their needs. There are dozens more examples, in fields spanning from financial services to health care—ranging in size from startup to Fortune 50.

BUTTONING IT UP

In scripted comedy, the "button" is the final bit, joke, or moment that "buttons up" what you've just seen. In a similar vein, each chapter will feature a "button list" of actionable applications of the concepts discussed. The final chapter will also offer another higher-level list broken out for employees, managers, and executives. You can skip to this, but it will make the most sense if you've read the preceding chapters.

We cannot promise it will be funny.

CHAPTER ONE

THINKING FUNNY

```
INT. EXECUTIVE'S OFFICE - DAY

From his Herman Miller "power throne," EXECUTIVE
is grilling MANAGER, perched nervously in her
uncomfortable chair.

                EXECUTIVE
        How did this happen?!?

                MANAGER
        It's a mystery! Employee efficiency is at
        an all-time high. Just look at this chart
        I spent *all last week* making.
```

2 ○ PITCH, SKETCH, LAUNCH

Manager shows Executive a chart labeled "Butts-in-Seats Time." The curve is rising steeply.

                EXECUTIVE
Oooh, that's impressive…but are you sure everyone is doing their best work?

                MANAGER
They have to be! I start every day by walking around, reciting your "10 Performance Commandments." Sometimes I sing it to different '90s hip-hop jams. I like to keep it fun.

                EXECUTIVE
Ha! That must be wonderful for morale!

                MANAGER
Without a doubt. All of their Mandatory Morale Reports awarded me an A. Except for Rockstar…who gave me an A+++.

                EXECUTIVE
So if we're getting so much done, how are those mouth-breathers at OtherCo still eating our lunch?

                MANAGER
Another mystery! I mean, we won "Best Booth" design at Castle Con. (looks down,

embarrassed) And had a much sexier "Booth Wench."

                    EXECUTIVE
I just worry sometimes that we're behind the Castle-Finding-Apps curve.

                    MANAGER
Impossible! After that enormously expensive— I mean, extensive—survey you commissioned three years ago? Your former college roommate Kevin's presentation had me in tears!

Executive leans back, smiles.

                    EXECUTIVE
Manager, I like how devoted you are. How can you bring out that kind of devotion in the developers?

                    MANAGER
Well, sorry to say, I'm fresh out of ideas. Can we get Kevin back?

                    EXECUTIVE
Unfortunately, our consultant budget is zeroed out…

                    MANAGER
What if we canceled Shy Guy's little "Creative Retreat" idea? Why do we need to

go to a castle to get creative? I can rent
a bouncy one for one-tenth of the price.

Executive jumps up, excited.

                EXECUTIVE
Aha, you see? The creativity is already
starting!

WITH KNOWLEDGE COMES POWER

Possibly that sketch above was entertaining, excruciating...and illuminating? We'd like to push even further and tell you it has the potential to be *liberating*.

That's right. As previously mentioned, we know firsthand how difficult organizational change can be. But what we propose in this book aims to take some of the burden and dread away, by helping you and your company to make changes that won't feel like new obligations—they'll feel like new freedoms.

I Don't Think That Word Means
What You Think It Means

Naturally, it all begins with developing a new mindset, which is what we'll tackle in this chapter. But before we can do that, we're going to take a cold, hard, diagnostic look at the words we use every day at work but rarely think twice about. When we look closer, we begin to wonder: Are we really saying what we mean and meaning what we say?

Productivity

For starters, let's look at one of the most beloved buzzwords of contemporary business: "productivity." To this we ask, is it "productive" to have workers, teams, units creating the most and fastest at lowest cost…if what they produce is so wide of the mark that it requires more time and resources plowed back in to correct it? Many business leaders today simply (and we would argue, simplistically) equate "productivity" with "being busier." But here's the problem: When everything that has to get produced requires an extraneous layer of support and maintenance, that takes away from their capacity to produce!

Efficiency

Another concept that business leaders like their managers to focus on is "efficiency." This is, of course, the idea that we should be able to make a thing in less time and with less waste or expenditure than we currently do. But how efficient is a process that moves quickly…and wrongly? What good is a process that's so efficient that you've eliminated any opportunity to pause, look around, and determine if you're even doing the right thing (like producing the "Mandatory Morale Report" in our sketch)?

Development

Why do we even refer to this process as "development," when it might rightly be called "manufacturing" or even "construction" of a piece of software? We use the term "develop" on purpose, because it suggests something that creatively evolves over time: like property, or a play, or—if you're of a certain age like the authors—photographs.

But are we doing a disservice by assuming that "development" is only of use when it yields something finished? What if we told you that there are better ways to value time that help reach closer alignment to actual client needs?

Deadlines and Timelines

Speaking of time, are we allocating it in the most value-yielding increments, or just the ones that happen to fit into other "boxes" in the work calendar? For example, what if we took the "dead" out of "deadline," made it less of a make-or-break "drop dead" benchmark, and more of a series of checkpoints? Perhaps this would give the project more opportunities to recover from injury, to regain its balance early and often, until it landed on the "delivery date" on a surer footing.

Similarly, a lot of stock is put in "timelines" as supposed guides to keeping everyone on track. But what if that track is veering in the wrong direction? We think better tracking comes from blowing up old ideas of how to time the various incremental stages of development—in favor of more nimble, utility-driven timing units.

And at the end of that process, what is a "deliverable," really, when it comes to the long-term relationship between company and client? Is it necessarily an actual concrete product, or is it the value/utility they sought out in hiring you? Another way this topic has come up in current critical business literature is: Should we focus more on *outputs*, or *outcomes*?

Failure

Likewise, what if we finally stopped throwing around the word "failure" as if it's a bad thing? Obviously, no one wants a failure at the end stage. ("Move fast and break things," anyone?) But what about embracing mid-development "failure" as a natural, constructive, and sometimes even indispensable step toward a more solid success? "Progress" and "regress" may seem like antonyms, but they both contain the Latin root of the word for "movement."

And to carry out that embrace, we'll have to develop a bigger appetite for sometimes wilder and wilder swings that, at first blush, seem deeply wrong. But when deliberately and thoughtfully built into our process, that larger amplitude for experimentation can yield a much greater body of knowledge, as well as even more valuable options for this product, and the next one.

Ambiguity

This mindset also calls for a different attitude toward "ambiguity." Not ambiguity about quality control or standards, of course. Rather, about what type of expectations we "need" the outcomes of our work to be met with. In brief, we need to start retraining our brains to accept that not everyone on Earth will love it—as long as the right people (and enough of them!) do. For example, both *Saturday Night Live* and the iPhone produce dramatically varying opinions in the public marketplace. And yet, both have a track record of delivering to a satisfied customer base that keeps coming back for more.

A NEW ATTITUDE: CONFRONTING THE NORMS

We all use these words, all the time, even if we don't fully understand what we're saying, because they have become company-cultural norms. But why is that, if they're not helping us? How can we change that? To create change, we'll have to look at where those norms come from—leadership. Well, rest assured, we're going after that too! For instance, we're going to call into question the usefulness of directives that come from above versus those that originate from within the team. We're going to spend a lot of time and ink on teams too, challenging you to rethink what makes them work best versus the sometimes inertial way they are thrown together. And lest you think this is *Lord of the Flies* time, we're not getting rid of structure. Instead, we're calling for new and better (and honestly, more fun) ways to use it as a tool, rather than an instrument of control.

So, if we really believe all these principles, these norms, the creative development process should have looked less like the sketch at the top of the chapter, and more like...

Well, actually, that "emergency" meeting shouldn't have happened at all! Everyone on the project would be aware of the timeline, not just the Executive. The Manager would be helping the Developers align their work with the Client and encouraging innovative-even-if-incomplete ideation, like Shy Guy offered. None of the "surprises" should have been so, because feasible, agreed-upon work goals would have been explicitly hashed out prior. Also, donuts? What a cliché.

To help illustrate these superior ideas and the other principles described above, let's turn our attention to an ingenious, and highly effective, model from the world of short-form creativity.

How We Like to Play

After the jarring events of September 11, 2001, America was in a state of paralysis. This was viscerally true in the creative community and entertainment business. At the time, Rob was working on a new TV topical comedy pilot, and days, weeks would go by with everyone asking each other, "How can we laugh at anything? Or ask anyone to laugh?"

With such sentiments permeating the arts, in November 2001, David Rodwin chose to reframe this rock bottom not as an end point—but rather something to bounce back up from. They felt the paralysis could be broken if artists were given the conditions, and charge, to *just make something short to show people, and don't think about anything else.*

Or as Rodwin described it, "How about we get ten writers and ten composers to write ten ten-minute musicals in ten days with ten performers?"

The result, Raw Impressions, was a multimedia arts nonprofit whose mission was "inspiring artists to be prolific with excellence." It brought together performers, writers, directors, producers, dancers, composers, and all production crafts; assembled them into teams of strangers; then gave each team a few parameters and a tightly structured deadline to produce a short film or live stage musical. These were showcased at events that ran in New York City and Los Angeles from 2001 to 2006. Over its tenure, Raw Impressions produced thirty-four film festivals, musical theater events, and variety shows—and attracted scores of top talent: Emmy/Tony/Oscar/MacArthur-winning stars, creators, directors, and more. All of them were busy working professionals doing this on the side from their "day jobs," for no pay, all of it yielding fully fledged, 100 percent original evenings of entertainment within one week.

But wait: Aren't "creatives," and Broadway/Hollywood people in particular, reported to be difficult, volatile, ego-driven? How could they be pulled together into highly functional teams on a microscopic timeline—and yet all unfailingly produce up to expectation, on schedule, and on (zero) budget?

Rob was a participant in two RIPfests (what the film wing was called), with his resultant short film "Pie Chi" getting accepted into seventeen film festivals and airing on Showtime. He remembers it as one of the creative highlights of his career. Here's how a typical cycle worked:

First, candidates were recruited by word of mouth from previous participants. Prospects had to be already working or have significant experience in a field of expertise. Some people were at the entry level and helped out as production assistants, but for the most part this was designed not as a woo-woo "artists' jam," but rather for people with an appropriate skill match. It was also crucial that any referrals consist only of people confidently felt to align with the project's culture, process, and expectations. More on those later.

Those who signed up for the current round would gather for a kickoff presentation, everyone in one room. Founders Rodwin, Bruce Kennedy, and Patrick Mellon (and later on, principals Chris Tiné and Erik Bryan Slavin) would explain the process and timeline. Then they would hand out and read aloud a manifesto called "How We Like to Play." This document went into excruciating detail about how Raw Impressions expected participants to navigate production hitches, externalities, interpersonal clashes, anxieties, and all the other "big feels" that can arise in such boiler room conditions.

According to Rodwin, teams were formed by the organizers via "pulling names out of a hat." Sounds a little anarchic, perhaps, but welcome to the tightrope of freedom and smart mechanics we're about to walk and hopefully lead you on. Each team was assigned a writer, a director, a handful of performers, a producer, an editor, a cinematographer, composers, and all the requisite craft professionals. They were also assigned a finite set of locations specially secured for the shoots. All participants were given the same small set of playful, creative prompts: a line that needed to be uttered at some point, a prop that had to be incorporated, a unifying theme or subgenre.

Then the teams would go off and work through the next steps. Writers would brainstorm their general first thoughts with the rest of the team, enough to get theoretical preproduction discussions going. Then the writers would have twenty-four hours to come up with their first (and really, close to only) draft. The producers and directors would take the baton next and work out a plan of action to film the scenes. For musical productions, writers would meet before the script deadline with composers to help them generate songs, exchanging works in progress as they evolved.

Then the production team had twenty-four to forty-eight hours to film or block everything needed. This was typically a rock-hard deadline, as it depended on locations being cleared only for permitted periods, often outside of business hours. Actors, hair and makeup, lighting and sound, and everyone else had been looped into the early production plans, and now had to be ready to move into action as soon as "Action!" was called. Rehearsals happened, but they had to be minimal and focused, and often right before shooting.

Then, with the film in the can, the director and writer would sit with the editor for a few days, hammering whatever had come in into shape, without the typical industry option of reshoots or supplemental photography. Rob recalls some incredible creative leaps that occurred during this period, transforming the shape and feel of what he'd originally envisioned in his script. In the case of live staged musical shows, the overall timeline was shorter since there was no postproduction, but the initial conditions were nearly identical.

Then, in an atmosphere of fevered excitement, all would gather again, as would an invited public audience, in a cinema or theater to watch all the finished productions.

Now here's the next leap we're going to take you on, in bringing the wisdom of endeavors like Raw Impressions into your own business: *Not all the products were winners.* Many were technically competent middle-of-the-roaders, some were outright stinkers, and a few were home runs. And that was not only fine, it was a core expectation. As Rodwin describes their mindset, "We weren't providing a product to an audience, but rather providing a process to artists."

Again, madness, right? How can a business with millions on the line and vicious competitors circling the waters possibly countenance "not sweating the product"?

We're here to make the argument that focusing first on the creators, and how they do their work, will get you to happier and more durable clients. But first, let's take a step back and look at the very savvy thinking underlying this process, and how you can learn to think this way too (and how your workforce will love to be there for it).

Founding Principles of Raw Impressions

One of the things that made Raw Impressions work so well is that they weren't trying to invent or reinvent a particular production model; they were simply removing the typical constraints of production (finances, producers, distribution/venue, marketing, ratings/box office). They just wanted to see what would happen if the focus of creation went back to creating and creators. There was no plan beyond trying this experiment once.

But the first Raw Impressions Musical Theater event was so beloved by both participants and audiences, soon there was a demand to try it again. Rodwin & Co. set about replicating the "experiment" as a working MO, and soon branched out into a film division operating along similar principles.

However, in its second outing, Raw Impressions hit a stumbling block. One of the key collaborators whose name was drawn from the hat proved unable to work well with others. They had the wrong attitude for the spirit of the thing, and given the tight time frame, that nearly proved fatal to their team's product.

So the Raw Impressions brain trust realized something crucial: They had assumed that personal referrals and verbal overviews would automatically bring in participants aligned with their ethos. But their fast-moving, impromptu problem-solving, highly team-focused approach—while appealing to many creatives—was so counterintuitive to conventional arts production, it had to be explicated. They had to *codify their expectations* and share them clearly with all participants—or as Rodwin memorably put it, "Treat everyone like aliens who know nothing."

Thus was born the "How We Like to Play" document. (See Appendix.) It's been handed out and read aloud at every Raw Impressions event since, and while there have naturally been complications, they haven't come from the people. In fact, it's likely that the type of people trained on such clear expectations were in a better frame of mind to overcome them.

Making Play Work for You

So, two takeaways from Raw Impressions' history, before we get into more specifics of what we like about their approach to mindset:

1. Figure out how to take away the barriers to creativity and replace them with a mindset of "Just give it a try, quickly, and at small-scale—and see what happens."
2. Figure out your precise expectations on how you want your people to work (or if you like, "play!")—and communicate them with stone-cold clarity.

WHY IT WORKS: ELEVEN MINDSET SHIFTS

At Sketch Development, we agree with a lot of the core principles that made Raw Impressions work the way we feel software companies should. We call them the "Eleven Mindset Shifts." Why eleven? Perhaps our comedy side inadvertently led us into a *This is Spinal Tap* homage, where our ideas go one step beyond. More likely, we didn't worry about conventional "even" or "round" numbers like five or ten, and just found eleven things that worked best—and isn't that what it's ultimately all about?

In other words, yes, we've reached that portion of the Business Book where we depart from narrative, research citation, and winning anecdote to...bring in the lists.

1. **Project vs. *Product*** —Software development companies, like many businesses, are built around the day-to-day notion of *project* management: "On time, on budget, within the assigned parameters." (We concede that that's one way to define "efficiency"—just not ours!) All those are important, but with attention focused on only those "rules of the road," it can become all too easy to lose sight of the actual destination: a valuable product.

2. **What vs. *Why*** —That said, it's not enough to even just focus on the thing—we have to step back and ask ourselves why someone has hired (or will pay) us to make it in the first place. What is the value of what we are making to the client or customer? To ensure that value is delivered, we must open our minds up to new ways of thinking about time, the vital as opposed to merely "traditional" parameters on creation (like requirements specifications), and of course, how we allocate budget not just to busy work (too often mistaken

for "productivity"), but to making things that work for those we are making them for.

3. **Complicated vs. *Complex*** —Many businesses, say automobile manufacturing, are *complicated*— they require an inordinate number of parts and mechanisms to be integrated, but once done, that process can be replicated infinitely to yield the same car. By contrast, software needs to create a new and different vehicle, preferably one the world has never driven before, every time. This type of work involves complexity—where outcomes can't be predetermined or guaranteed and your surroundings come into focus only as you get deeper into the work. This was part of the manifesto Raw Impressions taught to participants: Expect the unexpected, and prepare to adjust. And in Rob's case, allow your film to change shape in postproduction, to make it better.

4. **Execution vs. *Evolution*** —It's certainly impressive to *execute* well—to bring a solidly made and popular new product to market. But markets, tastes, and needs change, sometimes overnight. The company that's only built to create products aimed

at one consumer target will fall behind when that target, inevitably, moves. Instead, development needs to be done in such a way that market feedback can help it *evolve* (or *develop*), which means being able to move flexibly with the market.

5. **Working for vs. Working *with*** —We'll talk a lot more about teams in the next chapter. But the essential question here is, are teams putting something together *for* someone else outside the team, for whom it may or may not be right? Or as we prefer, is the person who knows what's needed most already part of that team, someone they're working *with*? In Hollywood and Broadway, a show may follow that first assembly-line approach, which can be slow and wasteful. But in Raw Impressions' unforgiving timeline, all stakeholders were working together and in near-constant communication of needs. And one week later, that paid off.

6. **Compliance vs. *Engagement*** —Again, we'll go deeper into management strategies later. But suffice it to say, we think you'll agree the best tools are those that get your developers personally

and fully *engaged* in what they're making, rather than merely *complying* with imposed limitations. Remember, Raw Impressions enticed plenty of highly credentialed creators to work under sometimes maddening conditions—because they got to enjoy stepping out of the "boxes" of their professional lives.

7. **Expertise vs. *Feedback***—We're all in favor of expertise. How could we write a book of advice and not be? However, too often expertise is treated as the end point of creative exploration. And as in Mindset Shift 4, that can be fatal when the market shifts into uncharted territory. Instead, we advocate building a system where feedback is an equally if not sometimes more valuable guide to how the product, and value, gets delivered. As Rob puts it, "If the audience ain't laughing, you haven't made comedy."

8. **Fixed vs. *Growth***—Ever dealt with someone trying to be the smartest one in the room, like our "Rockstar"? Or even worse, multiple people? What if your company's attitude was that there's no such thing, because the prevailing mandate was for

everyone to be constantly learning? According to the research of psychologist Carol Dweck, author of *Mindset*, this attitude makes for more powerful organizations and more empowered employees. And also, much less annoying meetings.

9. **Command vs. *Serve*** —This one's for you, managers and leaders. As with Mindset Shift 6, we are no longer talking about getting employees to just stay in their lane, we want them to burn rubber! And that requires a good pit crew. Okay, admittedly, we've broken the sports metaphor ban (we are nothing if not rule breakers!), but what this means is recognizing that the people you've brought on board deliver value—and are using your time and resources to facilitate that.

10. **Control vs. *Trust*** —Believe it or not, the healthiest way to view your employees is "unpredictable." Otherwise known as "human." Instead of treating code-makers as fungible robots, leaders must learn to guide through influence and trust. If thirty-four teams of strangers can trust each other to make films and shows that other strangers will watch, surely you and your day-in/

day-out colleagues can get comfortable with having each other's backs.

11. **Practices vs. *Principles*** —The problem with the term "best practices" is the unspoken question, "Best...for whom?" No two workplaces are the same. The best way to evolve (thanks, MS#6!) the practices that are best for yours is to first develop and agree on principles that best represent *your* workers and mission. Raw Impressions learned the hard way, but you don't have to—uncover your common expectations of each other and your workplace, and shout them from the rooftops.

GET SERIOUS:
HOW IT WORKS IN THE REAL WORLD

Alright, Serious Business Guy, we know what you're thinking: "Nice sentiments, John. Glad to hear that works for moonlighting short-form arts collectives. I've got an actual business to run!" Well, good news: So do two incredibly successful companies you've definitely heard of and almost certainly patronized—and they follow this same kind of mindset.

Make Money and Have Fun: W. L. GORE

You've probably heard of, or worn, Gore-Tex fabrics, a staple of outdoor wear. But its parent company, W. L. Gore & Associates, has an equally strong track record of innovation and market penetration in an astonishing variety of fields. Beyond outfitting expeditions to the South Pole and the top of Mount Everest (and space!), Gore has invented membranes used in hydrogen fuel cells. It has developed pioneering medical materials that have been implanted into thirteen million patients. It is responsible for the industry-standard products in both dental floss (Glide) *and* guitar strings (Elixir).

W. L. Gore has been in business since 1958 and had over $4.8 billion in revenue in 2022, making *Forbes* magazine's list of the top two hundred privately held companies in the US. And yet, according to Gary Hamel in *The Future of Management*, it's still a "big company that really does behave like a startup."[1] How?

A few examples set the tone of how Gore conceives of business—a.k.a., its mindset.

For starters, there was the vision founders Bill and Genevieve Gore had of what a successful company should

---

[1] Gary Hamel, *The Future of Management* (Boston: Harvard Business School Press, 2007).

be (as Hamel puts it): "a multiplier of human imagination." Already here we see shades of Raw Impressions' mission of "inspiring artists to be prolific with excellence."

Further, Gore's self-organizing teams (much, much more on this later!) are all driven by one central principle: "Make money and have fun." The first half should seem fairly obvious, but how rarely is the second one spoken aloud or written down in corporate life! Remember, it's about engagement, not compliance. An innovation company gets its most innovative work out of workers who love what they do. Also like "How We Like to Play," this *spells out* an expectation—rather than just hoping leadership will "be cool."

But let's go deeper into the wild (and wildly successful) way that Gore likes to play. At Gore & Associates, there are effectively no managers. A team is developed around any employee with an appealing idea and the proven diligence to follow through on it. As Hamel describes it:

> Gore is a marketplace for ideas, where product champions...compete for the discretionary time of the company's most talented individuals, and where associates eager to work on something new vie for the chance to join a promising project.

To quote one employee cited by Hamel, "If you call a meeting, and people show up, you're a leader."

This nonhierarchical approach rings numerous bells from our list of eleven: *trust* instead of control, working *with* not for, *serving* not commanding, and of course, a laser focus on *product*. It also has the benefit of creating passion-driven teams, who just plain want to be there, à la the ones that make Raw Impressions hum.

And don't forget, it also creates a lot of different things different markets want. And lots of money.

Story Is King: PIXAR

Granted, Pixar has only sent characters, not actual astronauts, into space. But if you've spent any time on Earth, you're pretty likely to have enjoyed one of its projects. Despite a few outlier stumbles, practically no other studio in Hollywood history has achieved such a consistent legacy of blockbusters, awards and nominations, and continued popular and critical reverence. Here, Rob draws our attention to the word "consistent." Almost every studio goes through a roller coaster of hits and misses; Pixar hits so often, and so deeply, it's not just something in the water. It's something in the brain fluids.

Fortunately, former Pixar president Ed Catmull opens a huge bay window into that thinking in his book *Creativity,*

Inc. In a chapter on trying to define Pixar's early identity (which can be another approach to mindset), he states one of the bedrock principles they landed on:

> "Story is King," by which we meant that we would let nothing—not the technology, not the merchandising possibilities—get in the way of our story. We took pride in the fact that reviewers talked mainly about the way that *Toy Story* made them feel and not about the computer wizardry that enabled us to get it up on the screen.[2]

Take a moment to let that sink in. Technology was and could still arguably be called one of Pixar's key differentiators. Merchandising has always been and will be the cash-hose that fuels entertainment properties, especially in family animation. But Catmull demotes both in favor of three of our mindset shifts: the *why* ("we're here to tell great stories"); the *product* (given that most studios religiously follow a "process" instead—buy Rob a drink and he'll regale you with horror stories all night); and *feedback* (the love, the feels).

---

[2] Ed Catmull and Amy Wallace, *Creativity, Inc.: Overcoming the Unseen Forces That Stand in the Way of True Inspiration* (New York: Random House, 2014).

Catmull's book is a gripping tale of how Pixar has repeatedly walked this talk. In one particularly harrowing tale from their uncertain early days, they threw the entire company and millions of man-hours into enormous risk to completely reconceive the story behind *Toy Story 2*—because the story wasn't working. The film delivered a whopping $400 million in profit—sounds like the risk paid off.

A second big takeaway from Pixar's successful mindset is to not only navigate but appreciate change and unpredictability. As Catmull says:

> Here's what we all know, deep down, even though we might wish it weren't true: Change is going to happen, whether we like it or not....Rather than fear randomness, I believe we can make choices to see it for what it is and to let it work for us. The unpredictable is the ground on which creativity occurs.

Catmull writes about how this thinking helped the company keep a "startup"-generated product focus even after its 2006 acquisition by Disney, and through a mind-bending series of radical changes...eventually becoming its beloved film *Up*. Let's put it this way: Originally, the film was about two brothers from a cloud castle who fall to Earth, where a tall bird helps them overcome their differences. Per Catmull: "Only two things survived from that original version: the tall bird and the title: *Up*."

Attentive readers may note that we are now solidly in the realm of embracing the *complex*: moving toward an undetermined outcome that only becomes clear as you get deeper into your work. From a personnel standpoint, a *growth* mindset explains how Pixar pivoted and pivoted its internal workings over and over again, rather than letting politics or traditions rule. And for a studio now still making crowd-pleasers thirty years after its first one, it's clear that Pixar has mastered *evolution*, even in the face of a dizzyingly fast-changing culture/marketplace.

Now it's time for your business to evolve too.

CHAPTER 1 BUTTON

Your mindset controls how your organization produces. If you see your work as cookie-cutter construction, you'll get cookie-cutter results. Your vocabulary, and your company's definitions of that vocabulary, is a window into your current mindset.

- Don't confuse productivity with being busy.
- Don't confuse efficiency with value delivery.

- If you see waste as a bad thing, you'll seek to eliminate it. If you recognize that waste as a necessary by-product of the process, you'll seek to mine it for value.
- Hierarchy is great for organizing people, but can get in the way of creating value.
- If you don't explicitly codify your expectations for how people should interact, your hierarchy will do it for you, with unintended consequences.

CHAPTER TWO

ASSEMBLING THE TROUPE

INT. BREAK ROOM - MORNING

The three DEVELOPERS are crowded around the table. ROCKSTAR ruffles through a bag of bagels.

              ROCKSTAR
      Ah yes, three everything extra-protein
      bagels just like I special-ordered!

              SHY GUY
      Is that…all you ordered for everyone?

                    ROCKSTAR
    It shows I'm focused on maximizing
    performance *and* flavor-tunity!

TEAM PLAYER fishes out a lone, pink bagel.

                  TEAM PLAYER
    It's fine. I'll just eat the *(shudders)*
    strawberry one…A little garlic and salt on
    strawberry never hurt anyone…

                    SHY GUY
    Maybe we could ask them to deliver…in
    separate bags?

                    ROCKSTAR
    Oh great! I've got this locked down, now
    you want to complicate things with Admin
    more?!

                    SHY GUY
    It's just that I have diverticulitis, so a
    stray poppy seed could send me to the ER…

                  TEAM PLAYER
    Oh come on, Shy. Chug some Pepto and walk
    it off!

                    ROCKSTAR
    And bonus! If you do go to the ER, I'll
    turn in your report. They'll be like,

"Doubly productive now?! Do you *ever* sleep?"

          TEAM PLAYER
Could I turn it in with you? I could really use a win after last week's presentation…

          ROCKSTAR
Nah.

          TEAM PLAYER
Alrighty then. *(sadly)* Go team…

TURF BATTLES AND TEARS

The morning bagel order never satisfies anyone. It always ends up with turf battles and tears. How is it that even a "fun" business like software can bring out the worst in its employees? For one simple reason:

Most companies today are better designed to preserve hierarchy than creativity.

Even if individuals make the mindset changes laid out in Chapter 1, nothing will change if the way the companies and teams themselves are set up isn't also questioned. Or to put it another way, corporations have a habit of evolving in ways that protect the corporation at the expense of the employee. And in this case, at the expense of the employee's creative brain.

Growing Conditions for Creativity

To be candid, the Fortune 500 is littered with many companies that operate in this way, successfully. They have mastered efficiency, competitiveness, profitability, market share, or some other combination of factors that lets them dominate without giving their full focus to innovation. But software is different. Software's competitive edge depends on solving a not-yet-envisioned-in-the-world problem in a never-before-invented way. In other words, it depends *entirely* on a creative solution (or really, several million compiled into one shipping box).

The most valuable resources in software are those brains, working their best and in their best way together. Software must be produced in conditions that elicit new forms of human creativity. And unfortunately, today's development environments are often not arranged in the best way to utilize this. Too often, a group of warm bodies with similar degrees but wildly divergent agendas are simply tossed together between four walls (or however many your funky startup building has). This has the same effect as tossing a pile of flour, oil, sugar, and eggs into an oven...and expecting cake.

In a similar vein, sometimes teams are assembled (and even hired) in a way that effectively reduces people to fungible skill numbers. A manager creating teams or hiring employees might incorrectly assume that, like a puzzle, if they line up the "correct" numbers and give them resources, the "right" product will emerge.

But this approach rarely works, because not just any motley crew will do! The ideal creative team does not arise from just assembling the most brilliant minds in one room. Rather, a better product will come from carefully curating a team with a mixture of styles, talents, and sensibilities.

What all these failed attempts at optimal team structure overlook is the fact that *software development* is creative. If this sounds repetitive, it's because the hierarchical approach is so ingrained in corporate culture that we need to shout it again and again. Software development is creative! There's an extra, indispensable, often invaluable human element to collaborative ideation that needs to be nurtured. And that requires a set of optimal growing conditions.

Suboptimal Conditions for Creativity (the Default)

But first, let's look at where we're coming from. What are the default conditions for software development today, and are they really suboptimal? And where do they come from?

To begin with, there are inherited institutional weaknesses in every company. One daunting challenge for creative software development is the bugaboo of "uniformity"—the pursuit of which causes creatives to unconsciously close off their own brains in favor of some perceived company "aesthetic" or "tradition." Even worse, it's a business weakness—if your firm isn't imaginative enough to get outside the box, one of your rivals will. This can be a bit of a thorny cultural problem, but we believe it can be addressed, and even reversed, by rethinking some structural issues.

And naturally, structural issues begin with the manager, supervisor, or executive. In the room, or outside, they basically have two choices: stimulate creative team energy, or shuffle it around. Most, not by intention but just by corporate organizational practices, end up doing the latter. Yes, there is genuine energy in the room, only... is it all really getting channeled to the whiteboard? Maybe an employee is hiding their best idea so they can save it for a superior, one-on-one. Or maybe the idea or solution wasn't right, but the employee pushes it too hard now, to impress a boss. Maybe someone decides their job will be more secure if they direct their meeting-room energy toward undermining someone else.

Of course, it's not possible to simply remove "office politics" from an office. But it is possible to consciously

separate power games from the creative process. Fear and status should be pulled out of brainstorming as much as possible. Not just because of their distortional effects on which ideas move forward versus which ones flounder—just think of the sheer amount of brainpower we *waste* in interoffice machinations! Can't we redirect those beautiful minds onto their best work? To crush the competition in software, we need all synapses on deck.

All of this raises the question of what role(s) a superior *should* play in a creative meeting. In software, there are certainly many approaches to this question that are not paying off. What that looks like, most of the time: A faraway C-suite boss, perhaps well-informed by solid ideas and practices, but whose mantras just don't fit what this specific team needs to do ("work smarter, not harder!")—and yet maintains an inflexible corporate stamp from above. And maybe an in-room supervisor whose primary goals for the meeting are "quiet, deadlines, deliverables, and a lunch order." These leaders seek **control**, and while that is often needed in many businesses, software development also needs a certain degree of **adventurousness** as well.

All the trouble doesn't come from those at the top, though. Those in the middle can create obstacles to a fully creative development meeting as well. What if the manager or supervisor privileges their own ideas (or those of someone with political loyalty)—but lacks the clout to win the team's buy-in? Or what if there's an aspirant—with some talent but not quite ready for the big time—trying to hijack the meeting into an early performance review? What if there are clear interpersonal power asymmetries, leading attention and seriousness to be doled out mostly according to status? Bad ideas getting protected just because of who uttered them? What if the "team" isn't even really a team per se, but rather an assemblage of just whichever top performers an executive could pull rank on and get onto his/her project?

Now take each of these dynamics, almost certainly more than one, and imagine how, as the process continues, they further cascade away creative energy. The faraway CEO's inspiring vision doesn't match what the dev team needs to make, so someone ends up leaching creative time crafting memos to appease that CEO. Meetings become a grind, as everyone goes through the motions of hashing out a product or feature they all know in their hearts isn't going to be "the one," but there was never a stone-cold tough assessment of. And if—make that, *when*—something goes south, just think how much of this precious, dwindling brain juice is getting inefficiently squeezed into redirecting, correcting, and sometimes just plain deflecting. Time, texts, emails are wasted trying to patch discrepancies between different company priorities and practices, none of which has anything to do with the product!

Somehow, despite all of these obstacles, the software does come together—maybe even very well—but no one on the team feels invested enough to spend an extra work-second on it.

What a headache! Sounds like there are almost too many ways a creative business environment can lose its spark. More often than not, it starts with the people.

Optimal Growing Conditions: A New Way of Thinking About Software Developers

What if we started thinking of software developers in a different way? Not just "code jockeys" or "product delivery specialists," but something more like this:

Unconventional craftsfolk doing their toughest and weirdest work, over a safety net composed of little more than the trust of their peers.

Yes we hear you, Serious Business Guy. It does sound a bit more like a circus than a tech company. Welcome to Sketch Development's radical reinvention of how to form and run a creative team that works at its best, individually and collectively.

In fact, the way we prefer it, it starts to sound less like a "team" and more like a "troupe." To understand the difference, let's turn to one of the most successful sketch comedy troupes of all time.

MOVING FROM TEAM TO TROUPE

In 1984, five comedians in Calgary and Toronto formed a sketch troupe called "The Kids in the Hall." The CBC gave them a TV series that lasted from 1989 to 1995, which was then rebroadcast by Comedy Central in reruns for...ever? Since then, they have produced a feature film and narrative series, launched a successful nationwide tour, and reunited for a new sketch series on Amazon. They are highly influential to an entire generation of comedians and performers, and are still beloved and avidly watched forty years since they formed. Rob teaches several of their sketches to his graduate students as models.

Here, we share with you some of the elements that made The Kids in the Hall troupe work so well for so long.

Dissatisfaction with the Status Quo

Now don't be fooled into thinking that they managed this incredibly long stretch of collaboration by being stereotypically Canadian polite non-feather-rufflers. Quite the opposite! A recent documentary about them is called *The Kids in the Hall: Comedy Punks*. And before they teamed up, each of the Kids was developing his own fiercely independent approach to comedy. According to Kid Mark McKinney, they only had one thing in common creatively: "We all started with different styles, but were all pushing back against homogenized comedy."

(Interestingly, the software development leaders who came together in 2001 to create the "Agile Manifesto" [see Appendix] also didn't agree on much, other than a "statement of principles" asserting what they were fighting back against in the industry.)

It was the Kids' dissatisfaction with the status quo that fueled some of their best teamwork. Says McKinney, "We were always coming to each other saying, how dark can you go, what's the meanest thing you can say? *We challenged each other to do our best work, like a competition.*"

We're going to return to that theme, but just let that process for a second. Imagine your employees coming into work each day, applying that kind of motivation to your product. Imagine what kinds of leaps and bounds it could lead to.

Diversity

The **diversity** of their different skills and approaches was an important source of creative vitality. As McKinney puts it, "This made things more interesting than divisive."

Dan Powell, an executive producer for numerous sketch shows whom we will meet in the next chapter, also praises diversity of perspectives as a value-add. He cites an example where two of his show's staffers would have vitriolic political disagreements, but some of the lines that emerged from those became memorable pieces of character dialogue on-screen.

This commitment to diversity also meant that troupe members with different strengths and weaknesses could pick up each other's slack as needed. When Rob was writing for *The Tonight Show* and CONAN, he and his peers would periodically experience "off days" or even "dry spells," but felt complete confidence that someone else from the team would come through that day. Admittedly, that "safety net" vibe can be tough to create. But if accomplished, it can spark an incredibly effective momentum for moving a constantly changing creative product forward.

Trust

However, it takes a degree of internal toughness as well. As McKinney memorably puts it: "We had a willingness to go along with different POVs, but also we really had to support each one. Four of us could say 'This sketch sucks,' but if the person who pitched and created it was passionate about it, we'd let it air."

Another word for that "internal toughness" is **trust**. Innovative, ever-changing work at the cutting edge simply cannot pull ahead of the competition unless there's a strong feeling of psychological safety within the creative team. They have to know they can flop, miserably, and show their worst sides to each other—yet still be fully supported in the march forward.

The first episode from the 2022 revival of *The Kids in the Hall* features two separate sketches where the fivesome are *completely in the nude*. Five men in their sixties.

If that doesn't say trust, we don't know what does.

Intriguingly, that episode—a premiere, reintroducing the Kids to a hyper-judgmental world—opens with an even bigger example of...um, "ballsiness." It depicts a yard sale where a customer purchases a copy of their 1996 feature film, *The Kids in the Hall: Brain Candy*...for one dollar.

The backstory here is that *Brain Candy* was a notorious commercial flop (though John and Rob will fight you to the death that it's an underrated creative triumph). The gag is that this dollar is the first dollar of profit that the movie has made since then.

That alone is an astonishingly forward-looking wearing of a failure on one's sleeve. We'll have a lot more to say about that in subsequent chapters. But our topic of interest here is what the Kids' experience with *Brain Candy* taught them (and can teach us) about team-building...*and* team-preserving.

THE LEVEL PLAYING FIELD (INSIDE KITH)

But first, let's take a quick step back. We've spoken so far of the Kids as if they met up in Canada one day, put aside all their differences, and then gelled forevermore. Far from it. But McKinney points to two factors that made them a successful working unit.

External Constraints

Imagine the external pressures of having to generate twenty-two episodes a season, for five years, on a tight schedule. Add to that the logistical complexity of a show made from, essentially, five to seven "mini shows" within it. The Kids lived this, which meant they didn't have time or bandwidth for personal frictions or unbridgeable creative differences to slow them down. They were young and hungry, and this was their "big break"—and it was turbocharging their careers. They had two choices: the grindstone, or the curb.

Internal Mechanics

Keeping it all together under such pressure meant that, internally, they had to develop their own mechanics for a coherent team. Whenever two or more members were fighting, one of the other two or three would step in immediately and mediate. They also fostered an anti-"alliance/clique" working pattern. In McKinney's words, "You just co-wrote with whoever was available or you felt drawn to." This bears some resemblance to W. L. Gore, as described in the last chapter. In both places, a good idea and its originator created its own management team, over and over again.

Solidarity (Not Hierarchy)

The complete absence of hierarchy, though, was the most important underlying mechanic, according to McKinney. "You want to have the feeling that you're on a level playing field with everyone," he says. "If one member is trying to pull rank, all the energy flows toward fighting that rather than toward the creativity."

Sustaining that energy flow can be challenging, as the troupe discovered when they came back together for the making of *Brain Candy*. At this point, the TV show had been long over, and the Kids—now beloved and bankable faces—had scattered to other, separate career opportunities. Dave Foley was the lead in NBC's hit sitcom *NewsRadio*. Scott Thompson had a recurring role on *The Larry Sanders Show*. McKinney himself joined the cast of *Saturday Night Live*.

And when they reassembled, the five did not work as well together. As McKinney explains it, having other jobs and prospects took away the urgency of that "make it or die" feeling they started out with. Also, the slightly looser production schedule of making a movie versus a weekly show removed the critical time pressure that had led to earlier pragmatism.

However, for McKinney, this was only half of it. The real issue was that the Kids had taken their sense of esprit de corps for granted, and just assumed they could jump back in and get to work with it. But they were all in different places now, personally, professionally, probably even to some extent comedically. They tried to jam all the puzzle pieces back together, but their edges were misaligned.

The result is not overtly evident in the resulting movie, or even a contributing factor to its box-office demise (talk to the folks in marketing about that one). But the process of making the film through "brute force collaboration" created a nearly five-year split in the group.

A few years later, though, the Kids were ready to try a live reunion tour. By now, says McKinney, they had absorbed the lesson of *Brain Candy*. They spent months just "playing," working out material new and old, growing to learn and relearn each other's sensibilities and techniques. From a traditional business point of view, this was "wasted," "unproductive" time and labor. From a sketch comedy one, it was essential to the product.

And the good news is, it did the trick. The tour was a success that led to follow-up tours in subsequent years, as well as the other projects alluded to.

No word yet on whether it will lead to further televised nudity.

THE SELF-ORGANIZING TEAM

Alright, Serious Business Guy, we hope you're starting to see this isn't quite a circus we're advocating. But perhaps a comedy troupe still seems light-years removed from the more sober-minded work groups of the corporate world.

Good news: It's not at all. What McKinney describes is, in essence, a concept known as the **self-organizing team**. It's been written up across business literature, including the *Harvard Business Review*—not exactly Clown College.

What is a self-organizing team? In his book *Management 3.0*, Jurgen Appelo describes the characteristics of one.[3] Here's our Sketchy spin on those:

---

3   Jurgen Appelo, *Management 3.0: Leading Agile Developers, Developing Agile Leaders* (Boston: Pearson Education, 2010).

- **Autonomy**: A self-organizing team makes its own plan and work decisions related to it, with members managing their own activities and taking ownership of how the objectives are met. In the next chapter, we'll see some variety in sketch comedy leadership styles. In some cases there are indeed central stars, executive producers, or other "main deciders." But even so, they defer to the team to do its internal work its own way first.

- **Collaboration**: Sometimes collaboration is treated as an "optional" activity when something goes awry, or you need to yank something over the finish line. In a self-organizing team, appropriately to our area of interest, collaboration should be a feature, not a bug. Remember the Kids' approach: "Work[ing] with whoever was available or you were drawn to." Rob had a similar experience writing for *The Daily Show with Jon Stewart*: Every day's morning task immediately led to the formation of provisional work groups. In candor, he found collaborating more enjoyable with some than others—but the collaboration was the norm, and everyone snapped to it instinctively.

- **Shared Responsibility**: In self-organizing teams, every team member should feel comparable responsibility for the overall task. As a corollary, each member ideally takes responsibility for different pieces of the work. This is a powerful two-shot that organically bolsters both team buy-in and intrinsic motivation.

- **Shared Leadership**: It goes even deeper: Each member of a self-organizing team should expect to step up and become the leader on a given subtask or objective. This way the team gets the bonus of experiencing each member's unique skills. More to the point, though, it builds up that elusive fuel called "trust" that we saw with the Kids. That enables both super-bold creative risks and a team that moves as a unit, leaching away minimal energy to interpersonal differences.

- **Participatory Decision-Making**: Here's one of the radical breaks with hierarchical workplaces: The team makes decisions, not someone on the outside or above. Or, even if big-picture calls are made from elsewhere, the team has direct involvement in those decisions and how they are implemented.

This is how troupes like the Kids leveraged that precious diversity of perspectives that we discussed earlier. More voices equals better choices.

- **Adaptability**: This is one of the greatest strengths that a company with such teams brings to the marketplace. Business, especially knowledge-based business, is volatile and fast-changing. So are self-organizing teams. When new inputs and challenges arise, the team is not forced to try to chart a painful middle path between the new needs and the old diktats from the Home Office. Using the principles listed above, they collectively team their brains to recharting the course, as often as needed.

Keep Trying Things

One reason for the efficacy of self-organizing teams is that teams with the attributes mentioned above tend not to give up. They keep trying things. We can find an excellent example of this in designer/technologist/TED speaker Tom Wujec's "Marshmallow Tower Challenge."[4]

---

4   Tom Wujec, "Marshmallow Challenge," Tom Wujec, https://www.tomwujec.com/marshmallow-challenge.

In this activity, a team is given twenty sticks of spaghetti, one yard/meter of tape, one yard/meter of string, and one marshmallow. They have eighteen minutes to use these components to build the tallest possible tower with the marshmallow on top.

This challenge has been presented to all kinds of groups and ages around the world. There are some very entertaining YouTubes of different teams' relative successes and failures. But one particularly striking pattern emerges for our purposes: The best performers tend to be kindergarteners. And the worst—business executives!

Why is this? When you watch how a group (perhaps not a group used to working together, and definitely not one who has ever tried this before) tries to organize itself to tackle this, you'll likely see two wildly divergent approaches. Usually, the executives (and to be fair, many other accomplished adults, including an embarrassingly large proportion of engineers!) tend to compete over who will be in charge, whose ideas will predominate, who has to fall in line with whom. To McKinney's point above, too much of the energy is flowing toward negotiating and fighting power dynamics.

Or, the team might act more like the kindergarteners—themselves quite new to the institutional authority structure of school, not to mention inherently low status—having no illusions about being in charge or proving themselves the best on the team. They just build! They keep trying things, whatever anyone suggests, with the winner always being "the best idea." Just like at the W. L. Gore company.

In a word, they self-organize to the task, rather than using a task as a pretext to bolster or change their own organizational standing. They have intuited the wisdom of all good managers: The most effective team is the one that feels most empowered to try new things.

HOW TO SMASH THE HIERARCHY

Like many concepts in this book, this new perspective on hierarchy may come across as somewhat utopian, or at least impossible to achieve in the modern workplace. Gary Hamel and Bill Breen capture this frustration rather piquantly in *The Future of Management*:

> The traditionally minded manager is understandably disconcerted when confronted by the reality of an organization where power is disconnected from position—where you can't push decisions through just because you're perched higher up the ladder; where you don't have "direct reports" to command; where

your power erodes rapidly if no one wants to follow you; and where your credentials and intellectual superiority aren't acknowledged with the laurel wreath of a grand title. For most executives, the synchronization of power with a precisely calibrated scale of management titles and grades is one of the defining, and comforting, realities of managerial life.[5]

Even if your business is trying to let go of such power-clinging practices, it can still be hard to fully pivot into the type of nonhierarchical team-building that works best for creativity. So here are a few techniques we've found to smash those old ways of working:

Integrate Trust Practices into the Work

No, we don't mean those cringey "trust exercises" they make you do in cargo shorts at the retreat. Recall what Mark McKinney said about he and his fellow Kids trusting each individual member enough to put on the air a sketch that the other four hated—and how this led to unprecedented creative leaps.

---

[5] Gary Hamel, *The Future of Management* (Boston: Harvard Business School Press, 2007).

Diversify Your Team

What we specifically mean here is a diversity of skills, approaches, and backgrounds. No doubt, there is something enticing about the idea of "putting the best brains together in a room." The only problem is, if they're all the same kind of brain (or strictly speaking, professional specialty), this can end up limiting the creativity. Even worse, a group of people on essentially the same career track (at the same company!) is likely to be tempted into competing for the attention and approbation of management.

Re-Center Expectations

We saw in Chapter 1 how Raw Impressions benefited from codifying and communicating expectations. But what specific expectations are needed for an optimally creative team? John has created a list of these expectations for Sketch Development we call "The Sketch Handshake" (see Appendix). And one that can be truly powerful for transforming your team dynamic is this:

> Some days you're more servant than leader. Some days you're more leader than servant.

There's a lot to unpack there, but here's what it boils down to: You are not necessarily doing the same job in the same position/job title every day you come to work. Every day is a different challenge, so *you* need to be different as well. You can either spend the day building your resume or a marshmallow tower, but not both.

Also, "servant"? Can you imagine your employees willing to take on that moniker? Yes, if you as their boss are as well. Smashing hierarchy begins with those higher up in it.

Train for Complexity

Although "complex" and "complicated" are often regarded as synonyms, they describe two vastly different kinds of business (and two mindsets, as you may remember from our earlier discussion). For a schematic example, let's say a "complicated" business is building an airplane. Here, inputs are understood. They may be numerous and wide-ranging, but they're accessible within a fixed realm of expectations. Expertise oversees the process, guiding the outcome predictably.

By contrast, a *complex* field is more like flying that plane. Yes, a clear destination and set of tools exist. But beyond those, a dizzying array of factors can arise, altering the flight path literally and metaphorically. One cannot build a plane, path, or even business model that will account for all such scenarios. Rather, the challenge is to build (or rebuild) the type of pilot (or back here on Earth, group of software developers) who works well in an atmosphere of complete collegial trust, is constantly assessing and reassessing the most vital needs, *expects* the unforeseen, and is geared up to respond to it creatively, sometimes innovatively.

Move from Groupthink to Troupethink

Remember McKinney's tale about what The Kids in the Hall learned from the fractious filming of *Brain Candy*, and how to reknit themselves as a troupe for their live tour? Your workplace can unearth a gold mine of insight from that. No, we're not calling for software developers to spend months running through comedy games (though that would probably itself make a rather amusing sketch).

Rather, we're borrowing a page from psychologist Bruce Tuckman's four-step approach to creating or nurturing a team toward higher performance.[6] Instead of taking the team's coherence and compatibility for granted (as the Kids did on the film set), Tuckman argues that they need to be guided through these steps:

---

[6] "Tuckman's Stages of Group Development," Wikipedia, last modified February 24, 2025, 02:54, https://en.wikipedia.org/wiki/Tuckman%27s_stages_of_group_development.

Forming

When a new group of people are brought together (or brought together in this new work configuration), they will need information: who each of them are and what each brings to the table, how they will fit in, and mostly, what their expectations will be. And yes, this might be a good time for some of those cargo-shorts trust falls. But many would-be team-building managers simply stop there. The key here, for Tuckman, is that this process takes some *time*, just for this gelling to properly set in. It is not a waste of company hours—it is a foundational investment.

Storming

Here's where things get "lively." Interpersonal tensions rise, authority and expertise are challenged, even the team's mission can come under attack. *Let it happen*, says Tuckman. It's the price of putting diversity into the team, it shows employee investment in the work, and ultimately, it's *unavoidable*. So don't try to squash or ignore it—expect it and lead through it. In the next chapter, we will dive deep into how to rethink meetings and other collaborative processes to accommodate and build on this, bringing us to...

Norming

If you've handled a) and b) well, here's where it all comes together. Employees start to appreciate each other's different strengths, align on a shared sense of mission, and most vitally, create an ecosystem of constructive questioning, feedback, and redirection without friction—among themselves and management. The key here is to bolster the connections being made among the troupe, while continuing to find ways to integrate individuals further into the team's project.

Performing

Now it's time to reap the fruits. But not to get complacent. Honor accomplishments in a way that lifts up everyone, rather than separating out "rock stars." Communicate and celebrate specific milestones toward project goals. Take careful note of what was learned from steps a) through c), so past mistakes can be avoided, and new practices replicated.

The astute reader will note that this is a second layer of complexity, on top of that described in the prior section. Humans are nothing if not complex, and groups of them not connected by "natural" affinities (blood, ethnicity, nationality, creed, etc.) are exponentially more so.

So be it. Turn that from a challenge to a competitive advantage. Remember: *At the end of the day, no one is better* equipped to navigate a complex business environment than a team built for complexity.

GET SERIOUS:
THE POWER OF TEAM SELF-ORGANIZATION

Enabling Teams at Aflac

Fear not, Serious Business Guy: Your reality check is on its way to the table. Want to see how powerful the idea of team self-organization is? Try the furthest industry from sketch comedy that we can think of: *insurance*.

A few years ago, Aflac (yes, the "annoying duck" people) became concerned about its auditing department falling behind, with as many as fifteen hundred unfinished hours carrying over into the new year's workload each January. Even worse, internal demand for audit support was increasing, at a time when the company was not able to allocate it more resources. Aflac wanted the unit to do more with less, the wish of all business leaders.

Fortunately, sometimes wishes can come true—when the company is willing to take a fresh look at better ways for its people to work together.

Aflac had recently adopted some Agile principles to its IT department and reaped impressive results. Now it brought in experts to apply that kind of thinking to the auditing department.

Before: Employees on a Conveyor Belt

Previously, auditing at Aflac was done by a "pool" of auditors. Individuals were pulled out of the auditing pool to fill needs that arose from projects, then returned to the pool to work on whatever came up next. Everyone operated under the supervision of two managers, who in turn reported to a director. As a result, assignments came from above, like orders on a factory conveyor belt, and whoever was available, whenever they were, would endeavor to keep up with the flow. But with an increased demand and no flow-of-work principles guiding them, the result sometimes felt more like Lucy and Ethel trying to keep up in the chocolate factory.

After: Enabling Teams

But under an Agile-friendly reconception, managers and directors were replaced by a **scrum master**. If you're unfamiliar with the term, a scrum master is a moderately trained "team coach," excelling in communication and organizational skills. It could be a member of the team tapped to help run things, an IT professional, or someone else from the company or brought in from outside. But crucially, it is *not* a manager or executive. A scrum master does not hand out directives or police adherence to corporate mandates. What they *do* is enable the team to figure out how to do its best work.

In this case, the scrum master took responsibility for all the tasks outside of the work itself: planning and plan development, determining scope, running meetings, review and sign-off. They held daily meetings with the team, but this was not your garden-variety "performative" meeting (as we will explore in the next chapter). The (sometimes) unspoken question hanging over every meeting, "What have you gotten done?," was transformed into, "How can I help you get things done?" According to a Gartner case study:

> This daily meeting allow[ed] the Scrum Master to coach the team, address delays, escalate where needed and make real-time, risk-based scope changes.

Critically, scrum masters also engaged in and managed constant communication with the auditee (i.e., client), continuously weighing expectations and progress, so that there were no surprises or delays at the end of a cycle. The scale and frequency of client communication and assessment were decided by scrum master and team together, removing another source of uncertainty, artificial deadlines, and "black box mystery" from this foundational relationship of business.

Continuity of Experience
Aflac also replaced the "auditor pool" with pairs of senior and junior auditors, assigned to the same project from start to finish, rather than "called up" ad hoc and rushed to the ever-accelerating assembly line. This pairing enabled a continuity of experience around the project, as well as the flow of knowledge from senior to junior. It also created a continuous, information-rich senior-junior feedback loop that eliminated the junior's guesswork or false starts (or at least prevented false starts from becoming false middles and false ends.)

The main divergence from traditional leadership is that the scrum master works *within* the team, essentially as a peer. Their focus is to determine and enhance whatever practices get the best and most impactful work out of employees, not simply the most or the "cheapest."

Freed up from the constraints of the hierarchy—and appeasing a boss (whose role was becoming inherently unappeasable due to increasing demand)—the team members now had the tools to figure out the best way to solve problems and innovate on their own, and/or in closer collaboration. Also, the mandated junior-senior collaboration, in effect, forged a tight and lasting connection, comparable to the well-knit team mentality that enabled troupes like the Kids to consistently perform at top level.

And, apologies for the Hollywood-unfriendly spoiler: It worked! The next year, Aflac's audit count jumped from 40 to 60, and its unfinished project-hours plummeted from 1,500 to 75. Team morale improved substantially. And clients have consistently reported higher satisfaction, in particular with the more transparent, collaborative, and predictable ways that a self-organizing team carries out the work in sync with their customers.

HOW TO VALUE AND NURTURE TEAMS

And now, a few real-world business examples into this book, why not part the curtain and show that we walk our own talk?

Inviting the Client to the Team: Sketch Development

Sketch Development doesn't just value and nurture good creative teams—under John's direction, they've built their entire sales and delivery model around them. As with Gore and the Kids, Sketch starts with the customer's specific needs and creates the perfect team around it. And that team must be multifunctional, including specialists in coding, testing, designing, gathering and implementing feedback, and deploying.

In fact, Sketch is such a believer in the power of a properly aligned and prompted creative troupe, they do something that might sound even crazier than The Kids in the Hall's "Chicken Lady" sketch: *invite the client onto the team.*

In our next two chapters, we'll outline how Sketch builds and assesses a product differently than in conventional development. But for now, suffice it to say: Sketch wants the customer to witness how a team-based approach works firsthand, and more importantly, *needs* them to be part of it. As you'll see, this approach works best in response to **constant, small-scale reassessment** and **negative feedback** (you read that right!). And both of those are much more valuable when acquired straight from the source than by the guesswork usually taken (often at great wasted time and expense!).

Sketch is striving to create a new standard for creative work that fits customer needs like a glove, so that requires having their finger in there poking away. Even if they can't join full-time, Sketch invites them to join the huddle as often as possible.

This is not to say, however, that Sketch simply throws the doors wide open and believes that "the more the merrier." On the contrary. A Sketch customer once asked them to put a forty-five-person team together to build software for them. As John put it, "Forty-five people is not a team, it's a collection of task takers." Instead, Sketch reviewed their needs and suggested a team of four people. Eventually they reached a compromise at sixteen, but the team never grew larger than twelve. And Sketch delivered the software way ahead of schedule.

Eliminating Management Roles from Teams

Another way that Sketch follows the troupe model is to, as nearly as possible, eliminate management roles from intra-team processes. As with the principles listed above, the team has to rely on each other to continue delivering, and gets to make all its own implementation decisions. Sketch handles engagement management separately, outside the machinery of the team, to keep maximum flow on problem-solving, not power-appeasing.

Keeping It Fresh
(Giving Employees Personal Project Time)

Finally, Sketch believes in the potential and talents of its hires so much, it puts a premium on them "keeping it fresh." For two full days every month, Sketch proactively pulls its developers away from client work to focus on personal projects. This accomplishes two things: It maintains a sense of autonomy and mastery, and it helps build bonds between developers who may become teammates on a future engagement.

Less time spent on a project may mean the project comes out better? Yes, if it's a time of maximum engagement, revitalized inspiration, and actual brain-on-problem time, not just office face time.

We'll see in the next chapter how to turn those face-time rituals into supercharged creative incubators. Every day.

CHAPTER 2 BUTTON

Instead of trying to eliminate politics—a noble but futile quest—find ways to separate delivery teams from the *effects* of politics.

- Find the right balance for your teams between controlled and reckless. Neither end of the spectrum is valuable, but there's magic in the middle.
- Strive to diversify teams. Diversity of thought, of experiences, of skills. One of the key tools for harnessing complexity is diversity.
- Create a safety net within the teams where it's not only safe to fail, but encouraged. You're not

looking for your teams to deliver without failure, you're looking for them to deliver in the face of inevitable failure.
- Creativity flourishes within constraints. What constraints can you provide them? Frequent deadlines are often helpful.
- Teams don't simply become high-performing once they're assembled. It takes time for team performance to develop, much like the products they create.
- Don't worry so much about whose job it is to do what. Focus more on ensuring you have the right diversity on the team to complete the objective, and give the team space to allow a solution to emerge.
- Reward the team for good performance, not the individual. Above all, don't reward the heroes who saved the company from crises of their own making.
- When management is necessary, consider whether it's necessary *inside the team*.

Chapter Three

THE PITCH MEETING

INT. MEETING ROOM - DAY

MANAGER is with the three DEVELOPERS, who have just finished presenting.

          MANAGER
    Interesting. I'll kick it upstairs and let you know—

Executive barges into the room.

                    EXECUTIVE
No need! Upstairs is here! Now lookit: I just spoke to Client, and we need a slight tweak…

Manager secretly pours whiskey into her coffee.

                    EXECUTIVE
Now they want CASTLR to include a full, 3D-animated projection of what would happen to their castle during actual medieval warfare.

                    MANAGER
You're talking about…pivoting from document and geography management to… 3D-game-engine-battle simulation?

                    ROCKSTAR
I'm on it! I'll create an amazing cut scene that shows the end user crushing and enslaving their foes!

                    TEAM PLAYER
Ooh! Ooh! And riding a dragon! It so happens that *I* already had Engineering work up some fearsome dragon noises and FX…

                    ROCKSTAR
Behind my back? When I was trying to get them to put in a sponsorship banner on the castle flags?

MANAGER
I don't remember them asking about sponsorship flags?

SHY GUY
Maybe we could write some code to overlay the geographical data we've already accumulated onto actual battles that took place there historically…? But that might take some time…

EXECUTIVE
More time? With OtherCo releasing "Fortressizer" next week? Are you crazy?

SHY GUY
Yeah, sorry. Never mind.

MANAGER
(to Rockstar) How quickly could we make "Dragongram" a reality?

ROCKSTAR
Thursday, if we just dump it out there and just let Reddit tell us where the bugs are.

EXECUTIVE
Thursday beats next week! And dragons are cool!

THE MISALIGNMENT OF MEETINGS

Whoa! What just happened there? How did that meeting go so far off the rails?

Or maybe—as troubling as this might be to contemplate—what if that meeting did exactly what it was "supposed" to do? And the problem is with the rails themselves?

Consider this equally troubling question: Are today's meetings (unintentionally, but by force of habit and tradition) more about meeting the needs of *customer* or *company*?

Obviously, there's a bit of a false choice in there— happy customers are intrinsic to corporate success. But the contemporary *culture of meetings* does seem to lean more heavily toward shoring up the company's institutional needs/weaknesses, rather than creating something customers will love.

It's a systemic problem that must be addressed, which is why we're now moving away from interrogating norms for concepts (Chapter 1) and interrogating norms for team-building (Chapter 2). Next we'll be entering the space where the ideas and people come together (or don't)—the meeting room—to look at the symptoms of misaligned meetings.

Symptoms of Misaligned Meetings

Groupthink

In the 1979 film *Monty Python's Life of Brian* (this is a sketch comedy book, it can't all be HBR studies), a crowd of first-century Jerusalemites have gathered outside the window of Brian, the man whom they believe to be the Messiah. Brian's mother, however, disdains the crowd and accuses them of sheeplike conformity. The crowd, which effectively proves her point by speaking each of its dialogue lines in one unified voice, protests, "We're all individuals!" Then one lone, dissenting voice peeps up: "I'm not!"

The hard truth is, in business meetings and other creative-organizational settings, we'd all like to think we're the dissenting voice of reason. But we're more likely part of the unison-speaking mob.

And it's not necessarily our fault. There's a reason humans have survived such a punishing world for so long: because we stick together. We experience some of our most transcendent moments (a concert, a religious experience, a sports triumph/disaster) by being on the same page. Even though our military history suggests otherwise, human beings inherently *like* to get along.

That urge to conform gets an extra boost from some of the forces at work in a conventional meeting. There are clearly dominant individuals (bosses, managers) who control our resources (paycheck) and status (position), and the incentive becomes strong, often insurmountable, to align with wherever that power is flowing. Even if it's not straight-up suck-uppery to the Big Dogs, agreeing with the group itself can be an overwhelmingly irresistible temptation.

By contrast, dissent incurs a cost. It takes a lot of personal confidence, earned workplace capital, and willingness to endure annoyance and pushback (and maybe even worse!).

But what if we could drop the threshold to speaking up, and break the stranglehold that "the group" has on its members' creativity?

Committee and Compromise Mentality

John and Rob are both husbands and parents, and do not for one moment dispute the value of compromise, or working out "the best messy solution for now." We also sit on many committees, of varying usefulness, but clearly some do serve critical functions.

However, we also both recognize that the "committee/compromise" mentality is not necessarily the best approach to solving *creative* challenges, especially the types that are faced in software. This is for two main reasons:

1. "Pretty pretty pretty pretty good" may work for Larry David, but in coding, either it works or it doesn't. And not merely at the heuristic level—just because the code functions properly does *not* mean it necessarily does what the customer hired you for.

2. Software is cutthroat innovation. Every company is competing to figure out a brand-new technological way to achieve something in the world. The winners are those who do it, not those who come up with something that "sounded cool in the room."

An effective software development team is not a democracy, nor a dictatorship. Instead, as in *Life of Brian*, it might be more helpful to think of one as a **tribe**—a collective of people who've aligned their values, yet are still willing to listen to one of their own go off on a crackpot vision—because he/she might just be onto something!

Or, of course, instead of "tribe," we might also say **troupe**...

MEETINGS ARE THE WORST

Now, of course, we do advocate getting things done. This tome would be headed to the Great Business Book Shredder in the Sky otherwise. But where we differ—and this is crucial—is on the question of what a *meeting* is actually supposed to "get done."

We all know the conventional view: A meeting is supposed to end with some kind of "deliverable." This could be a finalized blueprint, business plan, codified set of next action steps, or—in a creative field like software—an "idea" or "solution." In the most general terms, though, what is expected to come out of a meeting is (at least one) *decision*. Specifically, a decision that will govern the next allocation of time, resources, and manpower.

And yes, meetings do typically end with decisions made...because they have to, definitionally. Whatever the room has gotten to by meeting's end *is* the decision. But does this one highly finite timebox, with a group of widely varying perspectives and experiences and potentially limited information, necessarily have what it takes to reach *the best* decision? Or just *any* decision that gets us to the end of the agenda and on to the lunch order?

The better question, though, is: Is there a better way to run a creative meeting? Or at least the kinds of meetings that deliver meaningful advances in the product, not just the process?

And an even better (and more horrifying) question: What if your well-fed, well-caffeinated, well-compensated room full of the best and brightest gets together at its brain-peaking time...and comes up with a decision that sucks?

The problem is, there may be no way to know this within a meeting's environs. Eventually, the compiler, the testers, the focus groups, and others may find the flaws. But what if that's too late? Or at least, later in the process than you would have liked? Given what we've been exploring about the current (and to be fair, widespread and old-school) culture of meetings, there is little incentive to buck a trend embraced by the room—and every reason to keep your mouth shut.

Reframing What Meetings Are For

But imagine if you could reframe the expectations of a meeting, away from "We have to have *something*," to "We have to *advance our ability to build what our customer wants*."

First, you'd have to open up a little safe space—no, make that a huge one—for, lacking a better term, "acceptable suckitude." You'd have to figure out how to get your best people psyched up and fully prepared to march into a meeting of their peers (and reports)...and bomb! Repeatedly. Every day. Sometimes many times a day.

Well, strap on your Kevlar—we're jumping into the foxhole, with the best and funniest, to see what that looks like.

IN A SKETCH MEETING YOU CAN…

Sit Anywhere

The first thing you notice when walking into a late-night/sketch comedy pitch meeting is: *It's not clear where to sit.*

And that's the way it should be. Just grab a seat, literally anywhere. Rob's berth of choice was a beanbag, a few of which were often to be found in such rooms. But this isn't a military lineup, it's a potluck. The only position with any fixity is, at least one person (admin or team) sitting centrally enough to hear everyone, with a note-taking device and fast fingers. But otherwise, before word one has been spoken, we're already breaking up hierarchy with our butts.

Start Anywhere

Once all are settled and well-snacked (another pro tip—do not put creative meetings in tension with appetites and mealtimes!), the head writer (a lot more on this job title later) kicks things off. How? Sometimes just by calling on the person to their left or right, with "You got anything?"

Pitch Anything

Proceeding clockwise or counterclockwise (or in some cases, led by individual initiative), each participant proceeds to pitch, usually one to three ideas at various levels of development...or, as often as not, "not-quite-full-development." Indeed, one of the most common prefaces to a comedy pitch is, "This isn't it, but..." or, when things get feisty, "Okay this totally sucks but..." Rob and a former writing partner used to have an acronym they would communicate with each other through: "NTBSLI" (Not This But Something Like It).

Now this may sound like a cop-out or a slovenly work standard, but it's quite the opposite. In comedy, one of the most vital building blocks for a great idea is a bad idea, or at least an incomplete one that someone else can build on. This is in fact precisely how a good writers' room creates its best work: One person brings the kernel of something funny, while others tease out its richest layers of potential, and together the room mind pieces together the most creative and surprising way to present it.

Even if someone did bring in a highly fleshed-out and specific concept and execution, the room would still put it into play, riffing and revising much of it on the fly into something different. Call it emergence or "the wisdom of crowds," but an unavoidable transformation process happens in a pitch meeting room, and it almost unfailingly makes the idea better.

Pan for Gold

But how does this work exactly? How do we spot the gold when it's been crusted over by dross?

Dino Stamatopoulos has written for many late-night/sketch shows, but one of his proudest credits is HBO's top-of-class *Mr. Show*. Stamatopoulos describes how *Mr. Show*'s creator/stars Bob Odenkirk and David Cross would run a meeting:

> When Bob and David heard a shit idea in the room, they tried really hard to get everyone thinking about, "What made you laugh about this?"—really digging into the funny part.

Stamatopoulos recounts an example of this process at work. In season three, a writer pitched a sketch about a highly lethal roller coaster called "The Devastator," which keeps killing passengers, yet the amusement park keeps it running. The writer's original pitch envisioned the sketch as a press conference, where a glib, chipper spokesman for the park keeps defending the increasingly indefensible atrocities that the media keep bringing up. Although inspired by a darkly funny area of human experience—how we sometimes let "recreational danger" go too far—the sketch, as pitched, fell flat in the room.

And recall, there was still an entire room full of others, with their one to three ideas to get through as well. But the progression did not move on to the next idea.

Instead, that's when the room got to work.

With no particular order of outburst, voices arose: Isn't a press conference kind of a "talky" setting for such a visceral, dynamic topic? What about moving the action *to* the amusement park? Next to The Devastator itself, where we hear the screams! What about changing the relatively one-note park spokesman to a live local news reporter, forced to comedically balance "local-newsperson-cheer" with interviewing people staggering off the ride with bandaged, bloody heads?

The resulting sketch is arguably one of the most memorable of the episode, if not the series. But it only achieved its final form by being both taken extremely seriously and also thoroughly dissected by the room. And remember, it started out, as Stamatopoulos approvingly characterized it, "a shit idea."

Trust the Room

Dan Powell, an executive producer-writer who has helped make iconic sketch shows like *Key & Peele* and *I Think You Should Leave*, agrees with the necessity of "trusting the room." He says a big group of fresh ears provides the most useful, initial visceral response (laughter! Or...not). But a room full of fellow creatives also provides immediate quality control, or as he puts it, "The more eyes on it, the better." This "collective eye" not only offers more avenues for improvement (as we saw with The Devastator sketch)—it provides a critically early safeguard against a comic idea "we've seen before." In the comedy world, patently derivative work is a death sentence. That principle holds for software or other innovation-centered endeavors: If it's already out there, we need to go in a different direction.

Of course, in both fields, it's common to explain what we have in mind by referring to what others have done or are doing. According to Jakob's Law: "Users spend most of their time in other apps, so they expect yours to work like all the other apps they already know." But the difference is, in comedy as in software, it's lethal to merely replicate—your pitch has to build something new atop the old familiar foundation.

Refine the Results

On some of Powell's shows, in fact, the specific power of a collaborative room is used not just to generate ideas, but to *refine* the sketches further down the line:

Everyone would pitch to the room, people would pitch on pitches, and Executive Producers would huddle on them and decide which were the most promising ones. Then they'd come back to the room with the most promising ideas and have the writer "skeleton out the structure," for other writers to help them work out the kinks.

Then, once drafts and rewrites have happened, the script is brought back to the room another time, to punch up the jokes (perhaps comparable to an extra session of debugging or beta testing). Once again, the power of a synced creative colleague beats even what the most brilliant comedic mind can dream up.

Hit Your Rhythm

A final unique feature of a fired-up creative room is that it develops a *rhythm*. According to José Arroyo, a veteran writer for Conan O'Brien's shows, when the pandemic hit and the writers' room went to Zoom, it completely threw the creative energy off. As Arroyo put it:

> On Zoom you have to wait your turn to come off mute, or else there's audio feedback. This created an artificial "stop and start" rhythm. Versus in the room, there's a faster-moving, often-stepping-on-each-other

"riffing" energy that builds on itself and leads to more creative/comedic breakthroughs, in my experience.

You may have heard of Mihaly Csikszentmihalyi's seminal work on how people can get into an optimally creative "flow state." Well, as sketch comedy and other writing rooms show, that kind of flow can be taken to the next level when achieved by a team.

Depend on the Decider

By this point Serious Business Guy may be in despair, wondering where the dividing line is between "fast-moving creative jam session" and "unproductively shouty cacophony." That's where the **Decider** comes in.

Ultimately, even the most comedically ingenious flow of ideas, riffs, and revisions needs direction. But depending on the show, the power politics can be quite different. Some sketch shows are built around a famous face or faces (*The Carol Burnett Show, Inside Amy Schumer*). Some are guided by a powerful behind-the-scenes creative personality (Lorne Michaels of *Saturday Night Live*). And sometimes they arise from a preexisting, leaderless troupe (The Kids in the Hall).

Whatever the case, the important thing is that someone in the room is acting as the **filter** for the fast-moving creative rush. Importantly, filtering does not mean shutting down or cutting down "shit ideas." It means redirecting them, plumbing or replumbing them for that funny kernel, completely reframing, or just as often, building something newer and even more ridiculous on top of them.

This Decider is not necessarily the star or host of the show—though that person will sometimes pop in to the writers' room to partially collaborate. They're not necessarily executive producers or showrunners, either—the top-line people, who are often embroiled in working out one thousand other production issues.

Instead, the Decider is most often the head writer, essentially the writing staff's manager. However, the head writer differs from traditional managers in two important ways:

The head writer has *authority*. Yes, the head writer outranks the staff writers. Yes, they give the writers notes and directions. But, other than that, a good head writer does not tend to wield their power in the making of creative choices. Instead, they act more like a coach—jumping in to guide the team's emerging work into its best form.

The head writer also has *parity* with the other players in the room. A head writer must be someone who used to be a staff writer. Full stop. No rainmaker, outside consultant, or fresh hire from a different field could play this role. The writers' room is a highly specific creative problem-solving unit. The only people qualified to lead it are those who have faced those problems themselves.

As an outlier case, Powell recalls helping produce the sketch comedy series called *The Astronomy Club*. This was a troupe that had formed on its own and was performing live—and then got a deal to put their show on TV. They had no particular lead or star and, like The Kids in the Hall, considered each other peers.

Powell says this group made a conscious choice to avoid hierarchical creative decision-making by using secret ballots for votes. And since he was a producer but not part of the troupe, he would sit in meetings and "offer up ideas, more like an interviewer than an executive."

Both aspects of that dynamic—a group voting in secret, and an extremely light-handed facilitator—sound like they make up a freewheeling environment that could make Serious Business Guy nervous.

Instead, as we're about to demonstrate, they should excite you with the untapped possibilities of your creative pool.

MAKING THE MOST OF MEETINGS

According to Conway's Law, "The way you've organized people is going to show up in the product."

So here's our question: *Do you want a product that just "goes through the motions," or one that excites customers with its creativity and ingenuity?*

Easy question. Challenging execution. But it's all within reach, with a little attention paid to a few matters:

Getting Set Up

Basically, when, where, and how long do you do creative meetings?

First, when? The short answer is, frequently and predictably. The point is to introduce a *habit*—that you are expected to regularly bring ideas and additive brain skills together with your peers, to bounce off each other. At CONAN, the sketch team would have a morning meeting. They didn't know what time it was, just that there would be one. Often there was also an additional ad hoc meeting, called at any time. The norm was that each morning would have at least two unscheduled, impromptu brainstorming sessions. This kept everyone's brains primed and the creative fires stoked. The key is the *routinization of creative exploration*, and the not only acceptable but completely normalized "tentativeness" of results that emerge from meetings. This takes the pressure off "bad" ideas, and lets all ideas flow—the best way for the good ones to find their way out.

Second, where? In a world currently torn between the WFH habits of the pandemic and corporate desires to get everyone back in the office, our preferences lie solidly with the latter. Surprising, right? You'd expect the loosey-goosey comedy camp to favor hanging out at home in their sweats, but truthfully, the best collaboration only happens face-to-face, in the immediate energy of a room. As we heard from José Arroyo, a videoconference can capture some of this intensity, but also loses much of it.

Finally, how long? There's no set rule, of course, other than that creative meetings go on long enough to *meaningfully advance the product*. As in our discussion above, note that this does *not* mean the achievement of a "final decision" or "deliverable." It means that the meeting yields material, in one form or another, that innovates over what there was before and points toward what the next creative questions will be.

Shifting the Spotlight

We've all felt it. That clutch in our throat, that tension that flows into our body, that little buzz in our heads telling us to keep our amateur-hour ideas to ourselves.

That's right, the boss just walked into the meeting.

Now, we're not saying that all bosses impede creativity, or even that most of them do! But if they did, it would be fairly easy to understand why: Having the person who controls your professional and financial status judging your creative output may make you less likely to take risky creative chances, and more likely to say things you think they might like.

Another problem: Presenting ideas to an authority figure is a *Shark Tank*-style situation. It implies finality, and puts extra pressure on the ideas to be "ready for prime time," imminently actionable. If you've been following the thrust of this book, you know that that's unlikely to be the case. The best ideas don't just pop out—they have to be mined through collective brainwork. And time.

Time was key to how The Kids in the Hall made the jump from alternative regional Canadian stage acts to major comedy network stars. Before CBC even approached them, they spent many months figuring out their collaborations in tiny houses, where "no one came," says Mark McKinney. "And this complete lack of a spotlight for a while gave us space to grow and find our voice."

Rob can also attest to the value of "unsupervised" time. When he wrote for HBO's weekly *Dennis Miller Live*, the eponymous host came into the studio only for the one tape day. This freed up the writers to go imaginatively wild the rest of the week in what they came up with for Miller. And even though Conan O'Brien was a boundlessly creative and energizing presence in a room, whenever he walked into the writers' room as "the boss," the energy became more about riffing an impromptu comedy bit in the room...and away from pitching and ideating.

Now none of this is to say there should not be leadership in the room! But it should not be bottom-line, results-oriented—it should be allowing the process to play itself out. In software, we feel the ideal "decider" for the room should be the **product owner**. In theory, this might be the person responsible for the profitability of the product, or the current feature being worked on, or the vision behind the latest upgrade. What matters most is that they possess the fullest understanding of what problems need to be solved, combined with institutional knowledge of what's been tried or not. They also need to have both the clarity and confidence to throw an unformed challenge out to the room, ride herd through a wild stampede of responses, and see their way through to what helps improve the product and what doesn't.

In terms of day-to-day management, a product owner should also manage their backlog to *expect* some slack—so the team will have time and space to "play around." As we've been arguing pretty strenuously, that is not a luxury but a necessity to pioneering creative work. At John's company, two Fridays a month are allocated to this slack.

A full 10 percent of work days are invested in play time, and the investment has paid off. The time spent playing around has yielded a wealth of company improvements that wouldn't have come from any top-down directive.

And on the other side of the coin, having the light but unavoidable structure of the product owner refereeing the session gives the discussions a tough-minded focus. In particular, it avoids the danger of the "by committee" compromise that might come out of the room otherwise.

Finding the Gold

Remember how Dino Stamatopoulos described Bob Odenkirk and David Cross in the writers' room for *Mr. Show* listening to "shit ideas"? Instead of taking the easy (and fun!) path of joining the collective dunk on the poor hapless writer, these bosses dug deeper. They kept asking, "But what made you laugh about it?"

It may sound like a frivolous question, but it's exactly what should continuously be asked during development on a product: "What, if any, value is there to be unlocked here?" That question can cut two (helpful/Darwinian) ways:

1. Sometimes value isn't immediately apparent but resides beneath the surface. Don't be thrown off

by a middling pitch or presentation—dig deeper and free up yours and your developer's creative sides to see if it's there to be excavated, and the focus was on the wrong aspect of the product.

2. If there is none, no matter how you slice it, move on and don't stay encumbered to something whose value cannot be sussed out. Otherwise you're building up opportunity costs by the minute.

Now, Serious Business Guy might be skeptical that this type of focused leadership legitimately takes place in comedy rooms. You may imagine that they are nothing more than anarchic laugh-fests. But in fact, given the freewheeling nature of comedic ideas, especially in the extra-lunatic genre of sketch comedy, a well-made final product requires even more discipline. Rob spends months drilling his sketchwriting students on eliminating and streamlining extraneous ideas, tangents, funny but stagnant dialogue, etc.

Whether in comedy or software, a good meeting facilitator balances the unbottled liveliness of creative ideation with the laser point of product need.

Adjusting the Meeting

Like producing a sketch, developing a software product requires many steps, each of which may have categorically separate creative challenges. This means that a meeting could be devoted to only one type of directed collaborative imagination.

To draw on some examples from the writer's room, here are a few types of meeting ideas for developers:

1. *Brainstorm*—More like a hackathon, where the sky's the limit. This could easily also serve as an "icebreaker" predecessor meeting for a more focused one to follow.

2. *Fleshout*—Remember the NTBSLI (Not This But Something Like It) "half-baked" ideas? This is their moment to shine with an assist from others, or be burned to a crisp. The only thing they *cannot* do is stay locked away in developers' heads.

3. *Structure*—A sketch is a short, tight, highly unforgiving genre. Here's a tough meeting aimed at hammering out all the dead air, blind alleys, and lost laugh opportunities. Everyone takes a shot

at stress-testing the product, in a spirit of shared ownership and concern.

4. *Punchup*—This is every comedy brain in the room seeking ways to make every line, word, and delivery funnier. Comedy or not, in essence this is a round aimed solely at maximizing user experience and pleasure. Not a bad idea.

If all that work goes into something meant to last us five minutes, imagine what it takes to innovate software we'll want to use for months or years.

OVERCOMING COMMON OBSTACLES IN MEETINGS

At Sketch Development, John and crew take such "comedy"-esque principles seriously, and apply them directly to how they run meetings with clients. Here are a few examples of how the Sketch way in software slays the giants that can impede a truly "productive" creative session:

The "Nerves" Problem

One big difference between comedy writers and software developers is—to generalize, of course—the former tend to be extroverts, the latter introverts. What you want is a room full of brilliant people "riffing," fearless of failure. But with coders and developers, that isn't always going to happen naturally. Second, a development team may often be peopled with a variety of populations: people of different ages, backgrounds, and levels of experience. Such demographic separations could throw up yet one more barrier to a room that lets people feel fully comfortable letting down their guard and diving into "play" mode.

Sketch meetings overcome those kinds of barriers through a variety of techniques. They start with a few mandated minutes of almost-summer-camp-style icebreakers, guiding everyone to share about their favorite music, or foods, or something nobody in the room knows about you. They also break through the "fear of speaking in public" barrier by moving away from verbal pitches. Rather, a common Sketch approach is to give the room a prompt, then have everyone write their responses on sticky notes, which are then read aloud by a facilitator.

Visual ideating is another way to circumvent the "wrong words" anxiety. One exercise Sketch consultants regularly pull out is called "My Worst Nightmare." The room is broken into two or more subgroups, and each of them work collectively to draw images of their most harrowing work-mares. Then, the other groups have to guess what the drawings represent.

Serious Business Guy might turn up his nose at this as another, yes, summer-camp-type waste of Serious Business Time. But here's what we see happening: spontaneous collaboration, nonlinear thinking, and passionate (often downright joyful) engagement. And perhaps best of all, simply getting people to open up in a group setting about their fears is the perfect inoculation for them to move onto fearless pitches.

Room Politics

And that brings us back to one of the most common causes of such room fear: reaction from the boss. Afraid of looking like a fool in front of your manager or supervisor, you keep your half-baked, "NTBSLI" idea to yourself. Remember what we've said a few different ways by now: *Creative kernels that stay inside your developers' heads are incalculable value lost to your company.*

So, along the exact same lines as we saw with David Rodwin and Raw Impressions back in Chapter 1, Sketch is *loud and clear with expectations* on such roadblocks before the meeting even kicks off. A Sketch consultant will come and write on the whiteboard some of its most cherished "business creativity" principles:

- Check Your Titles at the Door
- Attack the Idea, Not the Person
- Always Assume Positive Intent

Sometimes they introduce playful (yet deadly serious) language about hunting HiPPOs—an acronym for the pernicious, unspoken-but-undeniable corporate habit: "Highest Paid Person's Opinion" (being held up above anyone else's).

Anarchy vs. Efficiency

So now we've successfully turned on that stubborn tap and your meetings are flowing like never before. How do we guide the flow so it's not just an out-of-control garden hose? One way is to use something called a **Liberating Structure**. Sketch Development did not invent these, and there are many of them besides this, but one of their favorites is called the "1-2-4-All."[7]

This format is particularly useful at trying to get a big group to come up with a macro priority—even something as "meta" as "What should we be having this meeting about?" It begins with each participant (the "1") writing their thoughts on one of those ubiquitously useful sticky notes. Then they are paired up with another (forming the

---

[7] "Liberating Structures," https://www.liberatingstructures.com.

"2"), and encouraged to "converge" their idea with the other's into a new, slightly more popular proposal. Then the pairs are themselves paired (the "4"), and also asked to try to converge. Finally, the four-person groups present to the entire room, and if all this is executed well, a strong "sense of the room" emerges.

Or maybe your style is to start without an agenda at all. You can do this with a meeting method called the "Lean Coffee." In a Lean Coffee meeting, the first step is for all participants to log what they want to talk about on sticky notes and post them on a board. Once everyone has submitted their topics, a facilitator curates them for duplicates, then asks everyone to vote. Each participant gets three votes that they can place on different topics, or they can double or triple up on a topic they care deeply about. The discussion then begins with the topic that received the most votes. A timer is set for five minutes, after which the group is polled to determine if they should continue talking. If the majority decides to continue (as indicated by a thumbs-up in the air), the discussion continues for two additional minutes. This continues until the majority is ready to move on to the next highest vote getter. This lasts the entirety of the meeting.

This format has worked so well for Sketch that they've built a freely available meeting tool, *leancafe.io*, as a digital representation of the format (hey, sometimes it helps to have some software people around).

Enough, Let's Move On

So those are some effective ways to start and guide a creative meeting. But what about ending the creative treasure hunt, or at least "knowing when to fold 'em"? The risk of a freed-up meeting such as we advocate is that it can also go off-road through either too much free flow, or conversely, too much perseveration on one issue by status-holders or their sycophants.

To cap it all off—this topic, this chapter, and most of all, such meetings—Sketch deploys what's called the "Elmo Card." It's a card with a picture of *Sesame Street*'s Elmo on it, given to everyone in the room. When someone feels that a topic has been overanalyzed, has been thoroughly mined for its greatest value, or is too far off of bigger priorities, they flash the card, whose acronym stands for "Enough, Let's Move On."

Indeed.

CHAPTER 3 BUTTON

Don't be afraid to pitch. Use techniques like NTBSLI (Not This But Something Like It) to convey the idea that you want to focus on the concept, not the content of your pitch. One of the most vital building blocks of a good idea is a bad idea. Pitch meetings should be designed to get the good and the bad. If you try to prevent the bad, you'll stifle the good.

- Be clear about who your Decider is. It helps if they've been in a delivery role in the past and faced the problems that the team faces.
- Pitch meetings work best when they occur on a cadence.
- Ideas that come from a pitch meeting are ephemeral. There's no need to capture those that don't make the cut.
- The best collaboration happens face-to-face, in the immediate energy of the room.
- Don't confuse the outcome of the pitch meeting as final. Dig deeper to find the real value.
- For teams that include introverts, level the playing field with sticky notes. Include a facilitator to give your meeting structure.

CHAPTER FOUR

ONLY LISTEN TO THE LAUGH

INT. EXECUTIVE'S OFFICE - DAY

EXECUTIVE is showing CLIENT a snazzy multimedia presentation (while MANAGER and the DEVELOPERS are covertly watching through the glass outside.)

          EXECUTIVE
  …and so, by synergizing augmented reality and AI with bespoke infrastructure, and state-of-the-art laser geosatellite synchronization—

ROCKSTAR
(*whispering to Manager*) The laser satellites were my idea.

SHY GUY
(*whispering*) Are they actually in orbit? Even after the launchpad accident?

TEAM PLAYER
(*to Shy Guy*) Shh!

EXECUTIVE
Our product is the first of its kind that can help users build the world's largest and most impregnable *sandcastle*!!!

CLIENT
What about finding real castles?

EXECUTIVE
We're…working on an upgrade…I've been told.

CLIENT
It's been nice knowing you.

Client gets up to leave. Executive freaks out, tries to stop them.

EXECUTIVE
Wait! Could I interest you in a Dragongram?

While not as extreme, versions of this scenario play out far too often at the unveiling of a new software product. It's like an unbridgeable gap appears between original intention and eventual outcome, over and over again.

But taking a step back, it's possible to see two main "cracks" in the system that ultimately become those gaps. And fortunately, it's much easier to spackle early than build the bridge later.

FOCUS ON USER VALUE

You might think the very first—and really, *only*—question guiding development is, "Who is this for and for what purpose?" Sadly, it is not. If you've been paying attention to our indictments of modern corporate culture, you might not be surprised to hear that other questions can often override, and often just plain bury, it. Questions like: "Will this make my boss/me/the company 'look good' in a presentational (as opposed to real-world operational) setting?" "Is this 'good enough' given the limited resources/personnel/timeline I was given?" "Are the specifications of this feature request detailed enough for me to provide a deliverable that will traceably satisfy the requirements?" "Is this idea 'cool'?"

Hey, you know what's the coolest feature a piece of software can have? *Working.*

In other words, for various reasons, developers and development have gotten oriented away from **user** and **user experience.** Products and features are looked at from thirty thousand feet (or from how they will "play" in the media marketplace), instead of on the ground, in the hands of the person who wants a very specific thing or things done, and wants to have a seamless experience accomplishing that.

A hodgepodge of different "values" now drives development, when there should really only be one: **direct, desired, accessible value to the user.**

Even if your company is on the right track to deliver something aligned with user value, it can still get derailed on the journey. This frequently happens due to the "crack" between product management and engineering often found in today's work-group arrangements. Put simply, the further apart they are (physically, interpersonally, and creatively), the greater the odds of mistranslations.

Sometimes the information gap is as simple as the inevitable game of telephone between two segregated departments. But this gap can be even more devastating than just a simple "misunderstanding," because gaps tend to get filled, and not always with the best materials. And engineers—under their own time constraints and pressure to deliver—are motivated to fill in the "info gap" themselves with a set of assumptions and experiences that may not align at all with the product managers (to say nothing of the end users).

As gaps get filled, assumptions—and mistakes—will get made. Very often, for example, the product team assumes (instead of inquiring or iterating through options) that the engineering "ask" is simpler, more feasible, or faster/cheaper to build than it is. As a result, the reality check from engineering comes back to slap product in the face too late in the process—necessitating wasteful course corrections. To quote Adam Sandler in *The Wedding Singer*: "This could have been brought to my attention *yesterday!!!*"

What often happens when these and other gaps widen, is that employees attempt to "gold-plate" over them. When engineering isn't delivering the precise thing that product has to sell upstairs, product may ask for some add-on features to disguise and distract. And turning to their reports, they may use elaborate sales language to accomplish the same thing: essentially, turd-polishing.

Would you want to drive over a bridge constructed of gold-plated turds?

LAUGH-TESTABILITY

Unless you picked up this book in the middle, you've probably guessed what we're about to say: "Consider how they do things differently in sketch comedy." As the title of this chapter suggests, any professional, broadcast/cinema/stage-worthy comedy product literally lives and dies by one constant, merciless, straightforward metric: *Is it making people laugh?*

That might sound obvious, but what's relevant to our comparison is, such productions build "laugh-testability" into every single stage of the process. Dino Stamatopoulos said of *Mr. Show* that this was the one and only driving question that shaped pitch meetings: "What made everyone laugh?"

Again, sounds like a no-brainer. In fact, this simple question calls upon three very specific brain activities: **investigation**, **optionality**, and **detachment**.

Investigation

This simply means that the aforementioned question doesn't stop at "Did this pitch/idea make everyone laugh?" It begins there. Stamatopoulos describes the next step as a collaborative writers' room effort to stress-test that "laughitude." Was it more than a one-joke idea? Could it move into different comedic areas, styles, and twists?

One of the lessons Rob drills into his sketchwriting students is that (by contrast to a sitcom or movie) a good sketch is a "smorgasbord" of different types of laughs that keep upping the ante. The laughs have to be strong and sustainable throughout all these "moves." More often than not, this is not the case at the pitch stage. But the classes he teaches are 90 percent workshop-based. The real work isn't always the idea, it's finding the comic potential and building a five-to-ten-minute "mini-movie" around it. That may not sound like a lot of time. But if you've ever found yourself nodding at a flat sketch jadedly, thinking, "Okay, we get the joke already"...what you are *not* doing is laughing.

Optionality

Sketch comedy brains are wired to throw as much "spaghetti" at the walls as possible, over and over again. "That didn't get a laugh? Fine. How about this?" On a typical day at CONAN, Rob would write between thirty-five and fifty monologue jokes (and was joined in this by three or four colleagues producing similar amounts)—for a total of around ten with the full expectation that, at most, one to three of any writer's jokes *might* be used. That was a norm, there and at other late-night shows he has worked for.

A similar calculation holds among those pitching sketches. Dan Powell breaks out some numbers that guided the process at *Inside Amy Schumer*:

> We needed fifty sketches for a ten-episode season. Seventy-five to a hundred sketches would make it through to the third draft. Thirty were no-brainers, but the remaining twenty were harder: Then these had to be weighed "against the season." (Was this sketch too similar to something else we're doing or have done? Is there another external issue or issues why we can't air it?)

Putting aside Rob's burning jealousy at that much higher usage ratio, you can see the pattern: *Most of the creative work one is doing in sketch comedy isn't going to make the cut. But you need a huge starting sample in order to find the gems that will.* There is no way to shortcut it. It seems strange to say this in print, but in seeking truly innovative and ad hoc creativity at scale (in both sketches and software), sometimes the secret to quality is quantity.

Detachment, Part 1: Don't Fall in Love

Stamatopoulos tells the story of how he once had an idea he found hilarious, whose pitch arose from him obtaining a $3,000 costume (usually, words suffice for a pitch, but bear with us). The room loved the overall gist and the costume...but they struggled with landing an idea that would reliably generate laughs throughout execution. They tried *twelve different iterations*, says Stamatopoulos, but none of them stuck. So they jettisoned it, and the show budget just ate the $3K. (Or, more likely, it eventually crept into the background of some other sketch.)

Mr. Show avoided the temptations of what John calls "falling in love with the solution instead of the problem." In other words, any investment of time or bandwidth or resources into developing an idea or feature can all too easily force further commitment to that path—even as it becomes increasingly unlikely to yield value. Both financial and psychological pressures gang up to pummel what should stay an open mind. And the longer this persists, the greater the opportunity cost of not exploring more fertile regions elsewhere. The analogue to "diminishing returns" is an audience cooling down and laughing less. In comedy, this is death.

Instead, to put this in ungainly comedy/economic terms again, the sketch-comedy approach is to build in an allowance for hundreds if not thousands of small "sunk costs." This accepted "budget" frees up the creative process to move, and keep moving, closer to its target—rather than plodding in the same, wrong direction.

As John likes to point out to his customers, the Manifesto for Agile Software Development (*agilemanifesto.org*) refers to a tendency to "follow a plan" so tenaciously, it gets in the way of "responding to change." "Change" in this context means the arrival of new information that didn't exist when the plan was hatched. Stamatopoulos's plan was to buy an expensive but gut-busting-looking costume. But eventually, the room chose not to stick to that plan, despite its steep initial investment, when new "information" came in that it wasn't generating laughs. In a conventional corporate setting, each of those twelve iterations might create even more pressure to stick with the plan. (Surely this is a funny costume and we're bound to get the right idea for it!) But it didn't, they detached, and nimbly moved on.

Doesn't that sound like something you'd like to see your company be able to do, multiple times a day, in a world where "new information" is *constantly* flooding in?

Detachment, Part 2: Kill Your Darlings

The corollary to "Did it make us laugh?" is "If it didn't, it's gone." Rob's sketchwriting class goes even further. To

Rob's standards, a sketch is ideally five minutes long. And in those five minutes, the audience has to be immediately immersed in a brand-new setting, introduced to brand-new characters, understand what the comic "friction" is going to be, then be taken on a high-and-escalating-energy ride (higher energy and comedically bolder than a sitcom or movie scene), through that aforementioned "smorgasboard" of varying styles of gag, then brought to a satisfying yet still surprising conclusion.

So in a sketch, every single piece of the script needs to be "load-bearing." No fat or dead air. In other words, a good sketchwriter needs to have a steel-clad stomach for *cutting cutting cutting*—what writers often refer to as "killing your darlings." If even comedy writers can be brutally functional with their material, surely Serious Business Guys can do the same. When Steve Jobs returned to Apple in 1997, one of the first things he did was drastically reduce the number of products Apple sold from their catalog. He famously drew a 2x2 grid on a whiteboard, filled it in with four products, and eliminated the rest, effectively reducing their product count by 70 percent. The impact of this move can't be understated. As a result of killing most of their darlings, Apple went from life support to consumer electronics giant.

How do you know what to cut? Here's a useful test from the field of eXtreme programming (XP), a framework that gave rise to the Agile movement: It's a concept called "YAGNI" (You Aren't Gonna Need It). Anything that hits the YAGNI mark needs to be eliminated. It's another way of demanding that every single feature in an innovative creative product plays a valuable and immediately identifiable function.

And if it doesn't? "Bye-bye, darling."

MAKE 'EM LAUGH: SECOND CITY

How do you discover what does make them laugh? It's not such a no-brainer after all. If you want to know, you have to have feedback. And yet another vaunted sketch producer we haven't even touched on yet has been ruthlessly focusing on customer feedback for decades, through their dedication to a very unfunny-sounding principle (which we'll define in more detail in the next chapter) called **empirical process control**.

The Second City in Chicago is a sixty-five-year-old comedy institution known for its style of improvisational theater, with locations in multiple cities that consistently play to packed houses. It was the original incubator for comedians like Joan Rivers, Dan Aykroyd, Martin Short, Chris Farley, Tina Fey, Steve Carell, and dozens more. SC is currently performing their 112th revue, "The Devil Is in the Detours." Like the 111 revues before, it's two acts of loosely scripted sketches, songs, and monologues. And very, *very* funny.

But don't be fooled by terms like "loosely scripted" or "improvisational." Second City has created an airtight system to harness the power of focused "end-user feedback" (i.e., audience laughter). When they first mount a new revue, they don't put the final product out in the world right away. Instead, they invite a live audience in and put on two acts of comedy for them. Then, at the end of the second act, the performers ask if anyone wants to stick around for a third act of improv.

For those who do, the cast takes suggestions from the audience and performs additional sketches made up on the spot based on those. But full disclosure: It's not 100 percent improvised. A lot of what they perform is another iteration of something they made up from a previous night that they wanted to improve, using the new suggestions as perturbations to help shape and refine it.

As an added benefit, note that the only audience members they have for this iteration are the ones that *wanted to stick around*. So in essence, they've primed their audience to become volunteer focus-groupers, thereby lowering the expectation for "final-show quality." This creates an environment ideal for market research. What they're really doing is producing the 113th revue in full view of their customers, constantly inventing, using the feedback from the audience as a guide to what should be included and what should be tossed.

Second City goes through this process every six months or so, creating two entirely original, "battle-hardened" shows per year. Comedy, culture, and trends have shifted massively over SC's six decades. But their commitment to closing the gap between development and customer satisfaction is unmovable.

WHAT IS VALUABLE TO YOUR USER?

A comedy "product" has a clear and undeniable value marker—laughs. We've seen how every stage of developing and even releasing new sketches is engineered to only proceed if laughs result. But how do you find that analogue in your business? (Not to mention ensure that all your processes are driving toward finding and maximizing it.)

Effective User Stories

One way is through what is sometimes called the **user story**. This is, essentially, a flexible narrative created around, or ideally with, the customer, aimed at surgically identifying *why and how they will use the product*. It is a far cry from a list of specs or requirements. Think of it as the difference between "We need a cloud-based word-processing app that runs on CRDT in AWS using CDK" and "We need to write a research-heavy book in real-time collaboration with coeditors in different countries."

What makes an effective, focused user story? Bill Wake—a leading voice in the early Agile software development movement—has identified a number of characteristics, and these are our favorites:[8]

1. **Negotiable**—Negotiations are rarely a one-and-done kind of thing, but rather a continuing conversation focused on goal rather than specifics... and one that can take many turns. This ongoing conversation is a creative one, and creativity requires "wiggle room." Or in comedy terms, "half-baked pitches." As we've seen, those are the essential stepping stones to progress. By contrast, demanding too much certainty too early in the process is a recipe for ending up with something that is certainly wrong.

---

[8] William Wake, "Invest in Good Stories and Smart Tasks," XP123 (blog), https://xp123.com/invest-in-good-stories-and-smart-tasks.

2. **Estimable**—On the other hand, we have to make decisions about whether or not to try out a new feature, and the cost of developing a feature is a significant factor. It would be helpful if the story was structured in a way that made it easier for everyone to say, "This feels similar in size to that other thing we built." Quick tip: If you want to make something more estimable, make it smaller.

3. **Testable**—Arguably one of the most practical parts of a user story, testability brings more discipline on two levels. First, as we've seen with Second City, frequent small-batch testing forces closer alignment between every step of the product and the customer. But it also clarifies whether the customer's story is indeed *actually* what they want! Better to find this out early, rather than after all that time and money has been wasted on developing the feature. If a user story doesn't have a straightforward test, you may need to go back and figure out how to rewrite the story so that it can be tested.

4. **Valuable**—And to come back to our question from the top of the chapter, *whose* value is served by this

product or feature? Is what you're coming up with going to be of actual, specific, and concrete value to the end user...or only the company's stock price and a few employees' promotions? A good user story should be able to sniff this out.

Jobs To Be Done

To get even more granular on that question of value, if you'll indulge us, we've got one more modern business concept to lay on you: **Jobs To Be Done**, or JTBD. According to JTBD theory, people don't buy your product—they hire your product to *get a specific job done*.

The most popular crystallization of this idea came decades ago from Harvard Professor of Marketing Theodore Levitt, who said, "People don't want to buy a quarter-inch drill. They want a quarter-inch hole!"

GET SERIOUS:
REALIGNING WITH CUSTOMER NEEDS

We'll round out this chapter by, ahem, "drilling" into two examples of how big corporations with already well-established customer bases used this wisdom to realign themselves better with actual customer needs.

What Customers Want: V8

If you're...let's say, around John or Rob's age...you might still hear in your head this red-colored eight-vegetable-filled juice's memorable catchphrase: "Hey! I could've had a V8!"

Now this was certainly effective, as the beverage's continued vitality shows. But one day, a brand manager decided to stress-test the product's value, with intriguing results. He consulted with innovation/growth guru Clayton Christensen, author of the seminal JTBD book, *Competing Against Luck*.

By taking a harder look at customer feedback on what the product did for them, a striking conclusion arose. Customers weren't buying V8 after asking themselves, "Which beverage should I choose to drink (or, per the tagline, face the bitter regret of having not chosen)?" Instead, they saw V8 as a more efficient way to "get their daily veggies" without any chopping, peeling, or cooking. Or, for the more squeamish, eating vegetables at all.

In short, nobody was demanding yet one more entrant in the already hyper-crowded beverage field. What they wanted was a more efficient health supplement. Armed with this knowledge, Campbell's (owner of the V8 line) moved their product from the beverage aisle to the produce section, next to the "solid" veggies it offered a more compelling alternative to. Sales quadrupled.

It's almost enough to make you say, "Hey! I could've taken a harder look at value!"

Why They Want It: McDonald's

Are we a little obsessed with food in this chapter? Perhaps. Or perhaps food offers one of the best test cases for how value can be taken for granted. ("Hey, if it tastes good, that's good enough, right?")

As proof that sometimes even world-beating behemoths stumble, this happened to McDonald's. Their problem seemed simple: They weren't selling as many milkshakes as they wanted to. So they attacked the problem on the level of "culinary value." They hired flavor and texture experts to try to improve both of those aspects. Arguably, they succeeded, and we've all been the beneficiary.

And yet, milkshake sales didn't budge.

So then, McDonald's turned to a JTBD approach. They poured research into a different and—on the face of it—odd-sounding question: "*Why* do McDonald's customers order milkshakes?"

What they learned was arresting—and incredibly valuable. Consistently, the most frequent purchasers of milkshakes are *drivers with long commutes*. Basically, this markedly slower-drinking beverage kept them occupied longer than soda, coffee, or juice.

McDonald's leaned into this appeal of the product, adjusted its marketing to celebrate it, and sales rose. On their way to ninety-nine billion more being served? Perhaps, at least for the branches with drive-throughs.

So now in this book, we've gone from thinking, to hiring, to ideating, to developing and testing product. All of which leave one final, essential question: How do we take all this and make it *sustainable*?

CHAPTER 4 BUTTON

Unless you're building software that is used by software developers, the developers are not the user. Always keep the actual users and buyers in mind when crafting features. A feature might sound like a great idea at first pitch, but resist the temptation to move immediately to development. Most good ideas shouldn't make it past the pitch.

- Don't fall in love with your solution. Even when you've developed 80 percent of it, leave open the possibility that there's a better way to solve the problem.
- Things change. Markets change. Preferences change. Priorities change. Your processes and attitudes toward development should embrace these truths.

- Bring as many customers as you can into your development process, as early as you can. Use their feedback as a guide to how and when to adapt, not as confirmation that you're on the right path.
- If you're looking for options to grow your product, seek to understand your customer better, not just your product.

CHAPTER FIVE

DEFINITELY READY FOR PRIME TIME

INT. EFFECTIVE SOFTWARE COMPANY

CLIENT, DEVELOPER, ENGINEER, MANAGER, and EXECUTIVE are all aligned and continue down a path of mutually beneficial, unexpected, and exciting collaboration.

(Not really funny, but *there's* a sketch Serious Business Guy can get behind!)

In a bit of a pivot (but we can handle those now, right?), we're going to conclude this book by switching from the short to the long. We've shown you many ways to retrain your eyes on smaller, function-clear units of directly testable value and how to reorient the supporting process (and personnel!) behind these around shorter, more frequent, focused but also free-to-fail creative sessions.

But how do we take a new model built on tinier bricks and expect to turn those into a skyscraper? In other words, how does all this lead to a more sustainable, scalable business over the long run?

Well, just ask an institution that's turning fifty this year, still occupies a central place in American culture, and even after five decades, is still able to pull in up to $300,000 for a thirty-second ad spot (that's, like, $36 million per hour!)—*Saturday Night Live*.

MADNESS, WITH A METHOD

Remember that image we presented you with in the Introduction: bleeding guest stars, leads getting into fistfights, no certainty from week to week of the show even staying on the air.

Before reading the book, you might be inclined to ask, "How did SNL go from *that* to the juggernaut we know it to be today?" But if you've read carefully, hopefully what you're starting to see is that this is the wrong question. What we should be asking is, "How did SNL become a fifty-year-plus juggernaut *precisely by doing that?*"

Because that is how it happened. And John—a longtime SNL superfan since childhood, now an adult working in software development—looked at the show's track record and began to think: *What a strange thing. This model of productivity that we have in our heads, it doesn't match up to what's going on here.*

So he decided to investigate. Technically, his "investigations" had begun in the mid-'80s, as an oddly young fan of the show. At age ten he was reading *Wired*, the John Belushi biography (which, for any of you parents, is not recommended reading for a child of that age).

John continued to keep a close eye on the show through the '90s, and decided to ramp up his investigation by infiltrating the system. He worked his way inside as an extra on the show—and learned a thing or two. (Buy him a beer and he'll tell you the story.)

His interest continued to grow so much that, when he started Sketch Development, he felt confident enough about this relationship between software delivery and sketch comedy that he decided to name his company after it.

So John resolved to really get his story straight. If there was a piece of SNL content out there, he consumed it: a book, a movie, an article, a biography, he took it in. And through all of that education, John learned a couple of things:

1. Yes, it is madness, but there is some method to it. So the cliché actually applies here.
2. That madness, that method, really only works *when the culture matches up to it.* If that culture doesn't exist, then that's all it is: madness.

So let us dive into some details here.

SNL: A SUSTAINABLE, SELF-ORGANIZED OPERATING STRUCTURE

Let's start by looking at how SNL put many of the techniques and attitudes we've discussed in the book into practice, and gave rise to a sustainable powerhouse. To do that, we need to understand more about *who* put these techniques and attitudes into practice.

Lorne Michaels was a comedian, who—as the founder and supervisor of the whole enterprise for all but five years of the last fifty—chose to act like less of a boss and more of a "player coach." In fact, when he started the show, Michaels was adamant that when the credits rolled, he would not be listed as the "executive producer," because executive producers were hierarchy, they were "the Man." (Hey, it was the '70s.) He wanted to make that absence of hierarchy feel like a vital part of the show. Recall our discussion in Chapter 3 about the head writer: someone who must rise from the ranks of the creatives rather than being imposed from without.

On the other side of that role, Michaels was also clearly and unquestionably what we'd call in Agileworld the **product owner**. He was not ruling the roost so much as minding product, process, and project; he was the keeper of the vision. As Michaels said in a recent *Hollywood Reporter* article, "At the end of the day, you just need somebody to say, 'This is what we're doing.'"[9] That is the role that he plays, just as a product owner does—because otherwise such a creative environment in a volatile marketplace can devolve into chaos.

To turn to the second leg of the self-organized-team stool, there's the **development team**. These are the

---

9   Lacey Rose, "SNL Trio Talk Trump Jokes, 2024 Election Comedy, Lorne Michaels' Future," *The Hollywood Reporter*, September 19, 2024, https://www.hollywoodreporter.com/tv/tv-features/snl-interview-trump-jokes-2024-election-lorne-michaels-future-1236005680.

people who are responsible for taking the product owner's vision and turning it into reality, something that other people can hang on to. In SNL's case, that's the writers and cast.

Finally, you've got the **scrum master**, who's not in the show and doesn't have any say in the vision. But without that person, the show does not go on. This is the stage manager for *Saturday Night Live*.

So that's SNL's unconsciously self-organized operating structure at the macro level, but they've also replicated it one organizational layer down. To illustrate with an example: SNL head writer Streeter Seidell once wrote a sketch called "The Tower of Terrors." If you're a fan of the show, you may remember it as the "David S. Pumpkins" sketch. It has been a breakout character and a viral sensation that gets referenced on social media every Halloween.

Seidell wrote the sketch. He got to cast it, and he cast guest host Tom Hanks. He even had the authority to tell Tom Hanks, "No, you can't refuse and punt this sketch to next week's host. It's going to be you." Seidell got to choose Beck Bennett, Kate McKinnon, and Kenan Thompson as cast members. He got to pick the costumes. He got to pick the set. The only thing he didn't get to pick was whether or not it showed up at 11:30.

Take a moment to savor the audacious (and extremely common) practice of such a money-driven, celebrity-driven, and tight-schedule-driven enterprise, comfortably allowing such major decision-making to take place at the sub-executive level. Ken Aymong, an SNL production manager for the show for three decades, crystallizes this audacious approach in the instruction he used to give to directors joining the show:

> You have to understand that you're going to be taking instruction from what you consider to be real neophytes in show business. You'll be expected to follow through on the instructions that they give you or the ideas that they have, because that's why we're here.[10]

At *Saturday Night Live*, each sketch is a cross-functional, self-organized team, empowered to decide what the details of the sketch are going to look like, constrained by the concept and format of the show. In order to build products that customers love, software teams require this same level of empowerment and self-organization.

---

10 James Andrew Miller and Tom Shales, *Live from New York: The Complete, Uncensored History of* Saturday Night Live *as Told by Its Stars, Writers, and Guests*, rev. ed. (New York: Little, Brown and Company, 2014).

Calming Creative Chaos

You might be thinking that sounds like total chaos, but it doesn't have to be. You can manage, and harness, the chaos using a popular business rubric, VUCA.

VUCA—*Volatility, Uncertainty, Complexity, and Ambiguity* In a software development environment, there is quite often **volatility** around requirements or priorities. Something super important has to be delivered tomorrow...but when you get to tomorrow, all of a sudden it can wait, or now absolutely has to be today! That's pretty much what an entire week is like at SNL.

In software, as in many businesses, there is also considerable **uncertainty** as to what the market is going to do. As a topical show, one of whose main cultural functions is to be the outlet that a nation turns to for its "take," SNL must react to the tumultuous news cycle every week. Even every day. For example, if our president tweets something on a Friday night, tens of millions of people are going to expect this show to respond. But it's literally something that cannot be planned ahead for.

Imagine being that nimble amid the **complexity** of managing production simultaneously on all three of SNL's live production stages (to say nothing of its virality-engine digital productions). Adding to that are the unintended consequences of one person getting in the way. When you're constantly trying to roll a set from one side of the stage to the other, there's complexity all over the place.

The same situation occurs in software development, where you'll frequently have scores of people managing one code base. Imagine if you had fifty to a hundred people trying to write a novel simultaneously; that's undoubtedly a lot of complexity.

And finally, we've got **ambiguity**. In software, it all goes back to that foundational question: *Is this feature that we're building going to satisfy our customers in production?* As we saw in Chapter 4, that's functionally the same as asking, *Is this sketch funny?* SNL wrestles with the same problems that software does—*but* they do it in a way that they can inspect and see how it's working (or not) throughout every step of the process.

EMPIRICAL PROCESS CONTROL

So how does SNL navigate VUCA? They manage it using, as we also saw with Second City in Chapter 4, **empirical process control**. Empirical process control is the polar opposite of what's termed **defined process control**—creating *the* plan, executing *that* plan, then getting predictable results from the execution.

Transparency

But under empirical process control, you don't create a plan and then execute. You create **transparency**, and multiple opportunities to **inspect and adapt**. As SNL and Second City, and more or less all the comedy shows we've referenced, do, you keep everything open and visible to everyone: the product itself *and* the stations of process. Transparency must apply to both, and that allows us to inspect what we've built. We can look at it critically, and if we find something wrong during inspection, we take action.

Now that might sound obvious, but when you think about the frequency with which this methodology requires you to go back in, *again*, and inspect and adapt, inspect and adapt, it becomes a lot harder—but also a lot more powerful. One of the rules of transparency John advocates at Sketch Development is:

If it's on the board, we're working on it.
If it's not on the board, don't work on it.

This helps eliminate extraneous, distracting problems like phantom work or pet projects. Instead, the watchword is "streamline." Get it all out there. Get everyone to agree. Radiate the information everywhere.

Transparency is also top of mind in SNL's approach to broadcasting the show. They frequently incorporate the live audience in some of their comedy. Before a commercial break, the cameras zoom out to reveal that the sketch they just did was set on a soundstage. It's an extra layer of transparency that brings the audience at home in to understand that, "Oh, this is a work in progress that we're looking at right now." This gives SNL permission to make a few mistakes along the way, which is indeed a proud part of its highly successful brand.

One example of SNL's robust culture of transparency came in the second season. Bill Murray was at a low point, and everybody wanted Chevy Chase to come back and replace him. Murray confided to Lorne that he didn't think he was succeeding on the show. Lorne's response was, "I think you should do a direct appeal." And with Michaels's blessing, Murray went on the air and literally pleaded with the audience, saying, "Please, think I'm funny." He seems to have done okay for himself from that moment on.

Inspection and Adaptation

Now let's shift from transparency to see how SNL enables **inspection and adaptation**. To do that, we take a deeper dive into a week in the life of SNL.

On Monday, the host shows up and finds that the entire show is a blank slate, that there is no script. This has been the same way, by design, for fifty years —even in the old days when there wasn't a half century of expectations to rest on. And to be fair, at first this could be a real problem. For example, Richard Pryor would not agree to be the host of the show until he saw the script. Ultimately, it took the writers visiting him at his home and becoming friends to prove to Pryor they were funny and get him to sign on.

So any given Monday at SNL begins with nothing on paper, and an NBC page traveling the halls yelling, "Pitch!" This is the signal for all the writers and the cast members to head to Lorne's office, where they rapid-fire out their pitch ideas to the week's guest host. It's a blend of real ideas, joke pitches, and suggestions from the host. At the end of the pitch meeting, somewhere in the neighborhood of a hundred ideas have been pitched. All pitches are cataloged. After everyone has left, the host, Lorne, and a few others review the pitches and throw out about half of them immediately. That leaves around fifty potential sketches for a show that can only broadcast about ten.

Then they announce to everyone which fiftyish ideas they're going to move forward with. Those are the ones that go to script. On Tuesday (usually an all-nighter), all fifty sketches get written. On Wednesday, everyone gathers for a read-through. In this first, early product-inspection session, they jam *everybody* they can into the room. Even the band leader.

The read-through lasts a long time because reading fifty sketches out loud takes as much time as it takes. After the last sketch has been read, Lorne and the host convene to select about twenty of their favorite sketches for staging. They've already reduced their options by 80 percent, but they're still going to produce twenty sketches—only half of which will get on air.

An NBC page once again traverses the halls, but this time yelling, "Picks!" This informs everybody which twenty sketches are moving into production. All of Thursday and Friday is a flurry of the crew building sets and the cast rehearsing.

At 8:30 p.m. on Saturday night comes the dress rehearsal. It is longer than the live show by about an hour (not surprising considering it has twice as much material!). Dress ends around 10:30, which leaves about an hour to review what happened during the rehearsal and make some decisions. Generally speaking, the sketches that generate the most laughter make it to the live show. But there are other reasons to cut a sketch. Perhaps an actor can't make the costume change from one sketch to the next, or maybe there are too many sketches about a particular topic. A variety of considerations have to be weighed to come up with the final list of sketches for the episode, and Lorne Michaels makes that decision.

Then it's 11:30, at which point the show is broadcast live nationwide.

THE *SNL* METHOD

Let's take a step back and look at this a bit more conceptually, at what we call the SNL Method. It all starts (on Monday) with a blue-sky free-for-all **Pitch**—good ideas, bad ideas, no holds barred, no filter. That list is immediately cut in half. The next step for the remainder is to **Sketch** something out. They create a rough skeleton of what this thing is going to look like, in the form of a script.

What they're doing is building a simplistic prototype of how the product might come out if they were to deliver it—then they eliminate half of those options and move right into "experimental" production. That's right, they knowingly engage the company's full resources into **Staging** fully developed scenes with twice as many options as what will be needed.

Then they **Test** those scenes by delivering them to actual customers (the dress rehearsal audience). And based on that direct feedback, they cut half more—and what remains is what they **Launch**.

What's interesting about this process is that as they progress from Pitch to Sketch to Stage to Test to Launch, they're simultaneously keeping a lot of options (200 percent!) open, progressively reducing those options, but consistently running them through multiple filters of feedback. Start with a small audience when Pitching, increase the audience size when Sketching (the read-through), grow the audience again when Staging (rehearsals), and bring in a bigger, external audience for the 8:30 Test show. And based on how they respond, you can feel confident in the quality of what gets Launched to the entire world.

What Doesn't Work (COVID Creation Fail)

An "exception to prove the rule" might be found in SNL's season forty-five, which happened during the COVID pandemic. After granting full respect/sympathy to the creative constraints on mounting a live comedy-variety show entirely filmed from guest and cast members' homes, the opening episode can still be instructive for our purposes. Tom Hanks (in a bit of a comedown from his David S. Pumpkins turn) delivered the opening monologue from his kitchen. Kate McKinnon did a "Ruth Bader Ginsburg Workout" from her bedroom. Kenan Thompson attempted to adapt his signature "What's Up With That" sketch over Zoom.

Long story short, it didn't work. By common critical consensus, the comedy fell flat. And don't just take our word for it. Here's what TIME magazine's Judy Berman had to say:

> The lows were predictable: too many stale *Tiger King* jokes. Too many sketches that had Gen Z influencer types blithely vlogging, doing DIY makeup tutorials and live-streaming video games in the time of coronavirus. An awkward split-screen edition of "Weekend Update," apparently recorded on Zoom with an audience listening in to provide laughter, was essentially a digest of the past month on Twitter. (Have you seen this doctor, Anthony Fauci?) Alec Baldwin called in as Donald Trump, mostly to complain about "nasty" questions. There were two separate Pete Davidson raps, at opposite ends of the show. Even McKinnon fell uncharacteristically flat with a Ruth Bader Ginsburg workout video that didn't work without the costume.[11]

---

11 Judy Berman, "'Live From Zoom': SNL Tries to Make Us Laugh—and Remember—the Pandemic," TIME, April 12, 2020, https://time.com/5819658/saturday-night-live-at-home-tom-hanks-review/.

And here's *Deadline Hollywood* weighing in on Weekend Update without the live audience format:

> Not exactly a bad joke, but torpedoed nonetheless by someone's inexplicable decision to have each quip met by the canned, lackluster chuckles of what sounded like three, maybe four people. The unintentional result was the human equivalent of crickets chirping.[12]

Even with some leniency granted to the highly restrictive format, we can see what's missing here: empirical process control. Even with incredibly hilarious talents putting their best funny bones forward, in the end this was a show comprised of sketches written in isolation, from home. At best, the sketches were likely approved by supervisors in an asynchronous, "notes"-driven process rather than the proven one: in-person, immediate-reaction, staged for larger and larger test audiences.

You can tell by taking a closer look at the complaints. The topical material felt tired because it was gleaned

---

[12] Dominic Patten, "TV Review: 'Saturday Night Live'-ish Tackles Coronavirus from Home for First Time Ever," *Deadline*, April 11, 2020, https://deadline.com/2020/04/tv-review-saturday-night-live-home-edition-snl-tom-hanks-larry-david-coronavirus-covid19-1202906456/.

from social media rather than stress-tested for freshness in a lively room. McKinnon didn't get a costume, putting too much pressure on the home format to deliver enough production value. Weekend Update's carefully audience-curated jokes didn't get their usual size of audience test.

Fortunately for everyone, the pandemic ended and SNL soon got back to doing what it does best—*how* it does it best.

What Does Work

For those not familiar with Agile concepts, Scrum is a framework that helps teams deliver in a model like this. Scrum is based on the idea of breaking your work into consistently short increments (say, two weeks) and then having something production-ready at the end of each increment. Or as Lorne Michaels perfectly summarized it:

> We don't go on because it's perfect. We go on because it's 11:30.

Another example of the SNL method—and this is really going to knock Serious Business Guy off his duff—is what they do with sets. In an interview with *The New York Times*, SNL's lead set designer from day one, Eugene Lee, said something shocking. He explained that at the end of every week, they completely disassemble every set—*even if they know that they're going to be doing another sketch that might use that set next week.* The philosophy was to never have to tell a comedian, "I'm sorry, that sounds really funny, but the set was not designed for you to do that." Laughs first, everything else second.

Suboptimize Locally, Optimize Globally

When John talks about this kind of practice with clients, he refers to it as **suboptimizing locally, to optimize globally**. Admittedly, not as funny-sounding a term as "The Coneheads." But it shines a light on the most key question for either enterprise: *What is the final unit of value?* It is an episode of *Saturday Night Live*—not a set on one of the sketches of *Saturday Night Live*.

John and his crew work with a lot of teams and departments who are so wrapped around the axle of being efficient that they don't realize that "perfect efficiency" in their department can in fact be creating massive productivity losses for the entire organization!

Hire for Strengths

Another window into SNL's ethos of continuous iterative improvement comes from outside of Studio 8H—way outside of 30 Rock even. But it drives straight to the point we explored in Chapter 2—hiring people who fit this culture is the bedrock of making its madness methodical.

Way before he joined the cast, Darrell Hammond was a stand-up comedian in Florida, and declared that he wasn't the funniest of all. But he made a deal with himself: As he drove his act from city to city in Florida, if he could focus on one thing to get better at each week, he could amass a pile of improvements. That adds up to fifty improvements in a year, two hundred fifty improvements over five years. Hammond's rationale was that while he might not be naturally funnier than the competition, he could outwork them. And as we know now, Hammond's results speak for themselves.

Have Frequent Dress Rehearsals

SNL's week also embodies a core concept that Sketch Development teaches to its customers who are writing software: *Don't automatically associate a deployment with a release.* A lot of software teams think, "As soon as I put this software we've written in a production environment, *that's* when our customers have access to it."

But Sketch, like SNL, challenges them to instead put things into production *as frequently as they need to*—and then let somebody else open the curtain to access those features at a later date. It's like the 8:30 dress rehearsal for the product or feature. Put it out there, deliver what you have. You don't have to open it up to the broadcast pipe that sends it to every television in America. But definitely get it out there in a controlled and eminently testable setting, until you are ready to share it with the world.

TRUE TEAM TOGETHERNESS

Let's go back one last time to that night in October of 1976.

There was poor Buck Henry, all clamped and bandaged up from his run-in with Belushi's samurai sword. But the show must go on, and he had more sketches to perform in...albeit with a giant bandage on his forehead.

Now even for a live, raucous (at that time anyway), semi-subversive late-night comedy show, that was a bit of a distraction. Possibly even a performance-degrader, in a fast-moving live environment. But an interesting thing happened.

Cast member Chevy Chase noticed the bandage (who couldn't, really?). And without telling anyone else that he was going to do this, he came out for his next sketch with Henry...wearing a giant bandage on *his* forehead! Then Jane Curtin saw what Chase was doing, and came out for her next sketch with one too. Soon, the rest of the cast was all performing sketches with bandages, with some even upping the ante to a fake sling for their arms.

When the cast members were interviewed years later, they all pointed to that moment in October 1976, where they looked around and realized, "Holy crap. We're a team, and we get each other." Without any external authority telling them what to do, they intuitively understood what their roles were as part of a troupe. In fact, this even brought the audience in, boosting customer loyalty.

This is a perfect example of the Tuckman stages of team development that we saw in Chapter 2: where teams move from **forming** to **storming** to **norming** to **performing**. The "bandage moment" was an example of a "storming" complication that threatened to throw things off. But instead, the opposite happened. The cast members chose (again, individually, as their company culture had empowered them to do) to turn that into a new "norm," and everyone else got on board. And from then on, that troupe reached a level of performance that solidified SNL as a mainstay on the American comedy scene.

A show whose title includes "Live" is explicitly built around the idea that you never know what will come up, but you have to be ready for it. The same really is true—though much less openly admitted—in a highly dynamic creative field like software. In both cases, leadership can either build what seems like a really tightly structured ship meant to survive choppy waters, or a nimble vessel whose captain and crew are all keeping an eye on conditions and ready to adjust to anything. But in our opinion, the first option is delusional, and could even lead to shipwreck.

And this discussion of leadership brings us full circle to Chapter 1 on mindset. In an interview, Lorne Michaels was asked if he was going to quit *Saturday Night Live* after fifty seasons—which is precisely where we are now. Michaels responded,

> Look, I'm going to keep doing this as long as I can, because at the end of the day, that's really all I am is the tiebreaker. I'm just the one to say, Okay, it's gonna be *this*, and that's what you need.[13]

In a strikingly humble way for such a famous and accomplished leader, Michaels's attitude demonstrates the deep value of a product owner. With so many balls in the air, you need somebody with the clear authority to say, it's going to be this sketch and not that sketch. Or, in your field, we're going to build that feature and not this feature.

---

13 Rose, "SNL Trio Talk Trump Jokes."

It may sound strange to say this in a motivational book, but SNL shows that a good creative leader is one who gives their people the opportunity to *not* shine. Because those stagger-steps, those error-ridden yet freely experimental forays, are the crucial framework from which improvement arises. If you're going to hide somebody, or something, in the back until perfection is achieved, you're squandering opportunity. You need a bad dress rehearsal to yield a good show. This is Product Management 101.

And at the end of the day—whether that day is Friday close of business, or Saturday night at 11:30, it all comes down to the team and the culture that sets their expectations. As Michaels learned over his fifty years—from John Belushi to Chris Farley to Pete Davidson—the only way it all works is if you are truly taking care of the cast. By establishing a culture of self-organizing brains not only liberated but encouraged to fail, as well as transparently inspectable, and relentlessly function-driven…you are not only making it possible to create things the world (and the market!) have never seen. You are creating an institution that can deliver it over and over again.

And that's nothing to laugh at.

CHAPTER SIX

THE BUTTON

As a summation of the Pitch, Sketch, Launch approach, we think it fitting to view what we've covered in this book through the lens of a retrospective. But not the stale, tired, finger-pointing or problem-denying ritual that sometimes passes for one in today's companies.

In fact, at Sketch Development, we like to say, "Retros are not an end point—they're a *starting point.*"

So how do you and your company *get started* on readapting to a more fluid, creativity-forward and priority-locked path of development? Here are the key learnings from each chapter:

MINDSET

Employees

1. **Adopt a Growth Mindset**: Embrace opportunities for learning and personal development. View challenges and mid-project setbacks as valuable learning experiences rather than failures.

2. **Say it!**: Keep the lines of communication open with your team. It's easier to stay aligned when you regularly share updates, challenges, and ideas.

3. **Focus on the Outcome**: Shift your focus from merely completing tasks to delivering value and achieving the desired outcomes.

4. **Embrace Agility**: Be open to adjusting your approach and methods as new information or feedback becomes available.

Managers

1. **Encourage Experimentation**: Create an environment in which employees feel safe to experiment and take risks without fear of failure.

2. **Facilitate Collaboration**: Diverse teams can generate more creative and effective solutions than silos. Find the walls between people and departments and tear them down!

3. **Show Them the Why**: Ensure everyone understands the goals and how their work contributes to the bigger picture.

4. **Make Feedback Your Most Powerful Tool**: Focus on providing feedback that helps employees and your product grow and improve.

Executives

1. **Champion a Value Delivery Culture**: Lead by example in prioritizing outcomes and value delivered over mere productivity. Make it clear that the goal is to create meaningful impact and get it in the hands of your customers.

2. **Support Continuous Learning**: Invest in training and development programs that encourage employees to continually upgrade their skills and knowledge.

3. **Build Autonomy into the System**: Prove to your teams that you trust them to make decisions and take ownership of their work. Empowering employees can lead to increased engagement and innovation.

4. **Throw Out Your Old Rulers**: Move beyond traditional metrics like output and efficiency. Focus on measuring the impact and value created by what you've released.

5. **Engage in the Process**: Participate actively in the development process. Provide timely feedback and insights to ensure the final product meets your needs and expectations.

HIRING/PERSONNEL

Employees

1. **Move from Shine to Share**: You want to really impress the brass, get promotions, and keep your job? Be a helpful part of the *team* that any truly innovative software product emerges from.

2. **Cone of Trust**: You were chosen to be in this room for a reason...and so were all your colleagues. It's impossible to know precisely what value you will bring to this product or feature, but your moment will almost certainly come. And so will someone else's.

3. **Be Flexible, Not Flexing**: When your work is on the cutting edge, you have to stay on your toes or you'll fall off the side. Expect things to change, be ready for it, and heck, learn to enjoy the ride.

4. **Own the Process**: In a Sketch-friendly workplace, your team members should be the ones figuring out the hows and whens of delivering your next unit of value. Contribute your hard-won experiential knowledge to those discussions.

Managers (and Recruiters!)

1. **Differences = Strength**: Everyone you hire should be smart and diligent. But if they all think the same way, why did you need all of them? Bring as wide a variety of brains as you can into the brain trust, and amazing things will emerge.

2. **Teams Are Made, Not Born**: Given that diversity, it's more likely than not your team may not gel at first, or even at second. Build in time and processes to build trust and get them on the same page.

3. **Collaboration over Hierarchy**: Every aspect of the workplace should be set up not to separate and enforce status, but rather to smash creative brains into each other at every possible opportunity.

4. **Expectations, Expressed**: Creative work attracts people who "think different." Make sure that doesn't translate into *working* differently from each other. State your expectations, and put them in writing. *Bonus*: Let your team generate their own expectations of each other...and you.

Executives

1. **Indulge Experimentation**: Software is by definition inventing a never-before-been-tried way of doing something. Remember how many times and how long it took Edison to get the right filament for the light bulb? Adopt the same patience to bring forth your hired brains' "light bulb moments."

2. **Be Status-Unconscious**: You're still guiding the ship, but a more empowered crew is not a mutiny. In fact, the more they lead their own creative process and let you focus on the big picture, the better for product and company.

3. **Value the Value-Bringers**: You'll never blow your rivals out of the water by getting something to market slightly faster, or more cheaply, if the product doesn't blow everyone away. Reimagine budgets and timetables as constraints that enhance creativity instead of stunting it.

4. **The Real Shareholders Are Your Customers**: Software users and critics are merciless. Don't show any mercy to the quality of the product. Your financial critics will get their payday when your software is the one people buy.

MEETINGS

Employees

1. **Let It Out**: An unformed idea is not wrong or bad. It's the foundation for a better one. Don't keep it trapped in your head for fear of looking

bad—the much scarier prospect is a potentially game-changing idea never being given a chance to flourish.

2. **Don't Fear the Fail**: Pitch the ridiculous, the so-wrong-but-maybe-just-maybe-there's-something-good-in-there, the "Not This But Something Like It." The more shots taken, the more likely a hit.

3. **Drill Till You Hit Gold**: Grab onto that ridiculous idea you just heard a colleague say and dig into what felt valuable buried inside. If another colleague tries that but hits bedrock, you take the pickax and go deeper.

4. **Rock Stars Are for Concerts**: You're in an orchestra now. Don't make the meeting about you or the snazziness of your presentation, make it about the idea.

Managers

1. **Should You Even Be in the Room?**: If you're a former developer or of a similar bent who got promoted to management, great. You know what's

called for and can facilitate. Were you brought in from above? Maybe sit this one out, and appoint a delegate instead.

2. **Early and Often**: Big meetings that try to cover too much, like big development processes that try to make too much, don't work. Pivot to shorter, more frequent development cycles to dream up the product in increments. Create a workplace cadence of "jumping in to solve a problem," and minds will stay juiced up all the time.

3. **Decide but Don't Dominate**: The brainstorm should be a wild and tumultuous ocean of creativity, but ultimately it needs a funnel. That's you.

4. **Foster Cooperation, Not Competition**: No teacher's pets, no protégés in training. You must value everyone you put in that room equally, and treat their contributions accordingly. Anything else shifts energy away from product to politics.

Executives

1. **Respect the Room**: Whatever wild and crazy things happen in that room may well make your

company a winner. Stay as far away from it as you can for as long as you can. Eight-hundred-pound gorillas do not inspire $800 billion ideas.

2. See previous rule.

PRODUCT

Employees

1. **Don't Hold Back**: Jokes and sketch pitches are a volume business. To achieve the same with a world-shaking software innovation, be prepared to throw *lots* of spaghetti at the whiteboard.

2. **But Don't Hold On**: Conversely, if something's not sticking and you've given it a hearty try, don't force it. You were hired for a brain that can keep coming up with more.

3. **No Guesswork**: What *you* think the product should be, or do, might be way off what the customer wants from it. Find out first: Pitch to your customer, sketch out an idea, stage a prototype.

4. **Bye-Bye, Bells and Whistles**: Throw out all discussions and pitches around things that do anything but make the product of direct, user-stated value. "When," "where," and even "how" pale beside the only question that matters: "why."

Managers

1. **User Story**: This is the seed the brain trust needs to ideate on-target. Get it straight as early and as simply as you can—a fuzzy story will not get any clearer on its own.

2. **Distinguish Iteration from Inertia**: We recognize this is a challenge, but it's one that you with your deep experience can learn. Become attuned to the difference between a process that is leading to more value—versus one that is just doubling down on a dead lead for other reasons.

3. **Refine As You Go**: Development is full of surprises for both company and customer. Keep the process nimble and the communication lines open with clients and market sources. That way, adjustments can be made before resources are wasted building something nobody wants. No surprises at the end!

4. **Test Early and Often**: Once you've nailed down testability from the user story, use it frequently, even at small scale and in rough form. Quality control that is baked into the process stays in control.

Executives

1. **Say No to Gold-Plating**: No matter how much you impress your shareholders, directors, media, they're not the ones buying the product. And they will definitely not be impressed if your impressive-sounding idea hits the market with a thud.

2. **Don't Get Gold-Plated Yourself**: People who spend a lot of time inside the corporate walls tend to use "lingo" to hide a turd-polish. Develop an allergy to that, and ask tougher questions that cut through it to the actual "Job to Be Done" by the product.

3. **Expand Your Appetite for "Waste"**: Appreciate the higher value brought to a product through more time spent exploring, iterating, testing, and reassessing. With the Sketch approach, all of that will save more time and resources in the long run.

4. **Don't Assume You Know Your Product**: The simple truth is, software is an especially fast-moving, and changing, field. The way you think about process and product should be as fluid and open to "new information" as that of your developers.

So to get back to where we started on this journey, Serious Business Guy, we hope that you can now appreciate how the Pitch, Sketch, Launch approach can be very serious business indeed.

And if any of this—no, maybe even *all* of it—sounds daunting, here's a little pep talk:

Remember why you got into software. Because it represents the ability of the human mind to create new, exciting things out of nothing but brain power. And you've collected some of the best users of that power in your company already. All you have to do is release it. You have a building (or campus, or empire) full of innovators. Why not direct some of that incredibly inventive energy inward?

And most of all, if this book-long assessment of the "business practices of sketch comedy" leaves you with nothing else, consider one last thing. People who work in sketch comedy come into work buzzing with energy and joy, almost never knowing what exactly they're going to be doing, but with one certain thought:

> Today I'm going to be part of unleashing something new, wanted, and pleasing into the world that will, in some small way, improve the human condition.

Imagine your employees, and managers, and executives—and *yourself*—walking into work with that in your heads every day.

Now imagine what might come out of your heads as a result.

APPENDIX

MANIFESTO FOR AGILE SOFTWARE DEVELOPMENT (AGILEMANIFESTO.ORG)

We are uncovering better ways of developing software by doing it and helping others do it. Through this work we have come to value:

- Individuals and interactions over processes and tools
- Working software over comprehensive documentation
- Customer collaboration over contract negotiation
- Responding to change over following a plan

That is, while there is value in the items on the right, we value the items on the left more.

Principles Behind the Agile Manifesto

We follow these principles:

- Our highest priority is to satisfy the customer through early and continuous delivery of valuable software.
- Welcome changing requirements, even late in development. Agile processes harness change for the customer's competitive advantage.
- Deliver working software frequently, from a couple of weeks to a couple of months, with a preference for the shorter timescale.
- Business people and developers must work together daily throughout the project.
- Build projects around motivated individuals. Give them the environment and support they need, and trust them to get the job done.
- The most efficient and effective method of conveying information to and within a development team is face-to-face conversation.
- Working software is the primary measure of progress.
- Agile processes promote sustainable development. The sponsors, developers, and users should be able to maintain a constant pace indefinitely.

- Continuous attention to technical excellence and good design enhances agility.
- Simplicity—the art of maximizing the amount of work not done—is essential.
- The best architectures, requirements, and designs emerge from self-organizing teams.
- At regular intervals, the team reflects on how to become more effective, then tunes and adjusts its behavior accordingly.

THE SKETCH HANDSHAKE

We're on a mission to improve the ways we work together. We're serious about our mission, and we expect you to be a significant contributor to it.

We choose not to go it alone. We expect you to help your teammates. That means, when it becomes apparent that something needs to be done to benefit the company or each other in some way, you'll raise your hand to be the person that does it from time to time.

We are active participants. Just jump in. We're all here for each other. A consequence to this approach is that it won't always be clear who's working for whom. That's okay. Some days you're more servant than leader. Some days you're more leader than servant.

We expect you to communicate with your teammates and customers. When your plans change, let them know. We believe the most efficient and effective form of communication is face-to-face. Be present in person when you can and on camera when you can't.

We expect you to share new ideas, big or small. We do not expect all of your ideas to be good ones. In order to discover the great ideas, we have to hear them all.

As you learn, we expect you to teach what you learn. Whether you want to write about it, speak about it, make videos about it, or however you want to spread your message, we expect that you'll turn what you're learning and discovering into an opportunity to start a conversation with our community. We do not expect you to speak at conferences *and* write blogs *and* attend meetups *and*...but we do expect you to do something.

We expect you to be honest and forthcoming when you don't know, or know how to do, something. We do not expect you to know everything, but we do expect you to want to learn.

You can expect us to help you find the answers when you need them. Don't expect us to give you the answers, especially if you haven't asked.

We expect you to let us know who you are. There is no such thing as who we want you to be. We want the real you. We have baseball coaches, tango dancers, powerlifters, military veterans, pastors, and actors working here. There's room for your individuality.

We expect that you'll approach your work with an Agile mindset and that you will work your hardest, but no more. We expect full-time employees to be regularly dedicated to Sketch forty hours per week (no more than forty hours). If your engagements don't sum up to a full week at times, we expect you to let others know what you're focused on with the remainder of that time and demo those efforts/accomplishments to your peers periodically.

To be your best, you have to recharge. You won't be limited to a predefined amount of time you're allowed to take off, but you will be expected to take at least five days of vacation in a row per year. You'll probably need some "me" time here and there too. If you're sick, please don't bring it around the people you're working with. We're all adults, and it's one less thing to have to manage in a spreadsheet.

You can expect us to provide you with space to get better at what internally motivates you. Curiosity and a desire to learn are at the core of who we are and what we do. Do not expect us to tell you (or even know) what those things are that you need to learn. You'll need to be vocal about this.

You can expect us to constantly look for opportunities for work that aligns with your strengths and desires.

Don't expect us to always have those opportunities available. Sometimes we'll ask you to work on stuff that isn't fun. Sometimes it'll last a while.

You can expect us to always look for ways to make this a better workplace. Please don't expect perfection in that regard.

We expect you to tell the truth. When things are good, when things are bad, and when things are sideways, truth and transparency are vital.

We operate under the Golden Rule: Treat others the way you want to be treated.

RIPFEST: HOW WE LIKE TO PLAY AND THINGS YOU SHOULD KNOW (V.5)

Please take the time to sit down and read this completely. Even the parts for other disciplines. It'll answer questions you didn't even know you have and set up the whole group on the same page.

I. **EVERYONE**
   A. THE SPIRIT (a.k.a. our corporate culture)
      1) Raw Impressions is committed to inspiring artists to be prolific with excellence. We also create opportunities for artists to take risks and work with new collaborators.

Lastly, we premiere the work we commission after supporting it through a speedy development process free from commercial pressures so our artists can just be artists.

2) Arrive with no expectations and be prepared to let go of attachments to any plans you make once we begin.

3) Raw Impressions is committed to supporting all participants in creating brilliant performances, inventive work, and exciting collaborations through an excellent process. However, keep in mind RIPFest is *raw*. Don't expect you can achieve perfection or realize exactly what you had envisioned. Make use of all obstacles put before you—planned or unplanned—so they become opportunities.

4) Your team is perfect the way it is. Another team has a choreographer and dancers. Yours is perfect without them. Your team has a DP with a Steadicam. Your team is perfect with that. It rains. Your shoot is perfect with rain. Even if it's not in the script.

5) There are continually last-minute changes to schedules, rehearsals, locations, script, personnel, etc. Please roll with the punches, persevere, and do the best you can. We want to do great work, have fun doing it, and put together some great films.
6) Everyone is working under extreme time pressure. Meet your deadlines. Start promptly and end on time. Our greatest commodity is time. Don't be late. On time means fifteen minutes early, warmed up, in position, and ready to begin.
7) If you are late, call your producer and give him/her an estimated time of arrival. No one cares why you're late, just apologize, say when you'll arrive and that it won't happen again.
8) Don't assume the producers know what you need. Make requests if something isn't provided for or mentioned. Ask for it as early as possible.
9) There are one hundred participants involved in this event. Every single one has specific requests of the producers. Please know that the producers will seriously

consider all requests, but not everyone can be accommodated at all times. Please trust that the producers are acting in the best interest of the whole group and all of the work.

10) Raw Impressions, Inc. is a not-for-profit 501(c)(3). All participants donate their time, energy, and expertise. We're committed to every person getting more out of their experience than they put in to it. And we ask everyone to put everything they can in.
11) No divas. Period. We don't have time for it.
12) Remember, it's just a film. No one's going to die.
13) Come to all the parties. We're determined to have fun, damn it!

B. COMMUNICATION
1) Check your email every morning during the RIPFest process. If you don't have regular email access, tell your producer ASAP and you will receive calls when there are updates.
2) Keep in communication. If you think something but aren't sure, please *ask*.

3) If *any* issue arises and you don't know who to talk to, call your producer to cut off any problem or concern before it grows. If the problem is with your producer, contact the executive producer. You can also call David 24/7.

C. OWNERSHIP and FUTURE USE
1) Raw Impressions, Inc. owns the copyright to all the films created under the auspices of RIPFest.
2) *You need to sign a letter of agreement before the end of the first meeting.* Before that first meeting, we will email you a copy of the agreement we need you to sign so you can peruse it. If you have any questions about the agreement, contact your producer immediately.
3) Actors: We are using the SAG Experimental Contract. You can include any part of the film you participated in as part of a reel to promote you and your work.
4) Directors: With the written permission from an authorized representative of Raw Impressions, Inc., directors

may submit their individual film to festivals. Raw Impressions Productions may pursue theatrical release, festival presentation, and broadcast in all mediums for some or all of the films created during RIPFest.

D. PROCESS
1) Teams will be announced at the first meeting. Each team will include at least: a project-producer, writer, director, composer, editor, director of photography, one to five actors and possibly a PA/AD.
2) Guidelines for creating the films will be given out at the first meeting. These guidelines are determined by the producers. The guidelines for RIPFest #1 were:
    - Create a five-to-ten-minute DV film tailored to the actors you are given.
    - Use the one indoor and one outdoor location you are assigned.
    - Theme: East vs. West.
    - Include an off-screen event.
    - Include a continuous thirty-second section of no dialogue.

3) We encourage collaboration in the creation process. Every team will find different ways to do this and create different boundaries. The entire team is encouraged to contribute during the brainstorming around the first team meeting after the launch. In the end, the writers choose what they will write about and how to incorporate people's ideas *and* the directors will help develop and dramaturg the script.

E. SCREENING

1) The producers will determine the screening order of the films.
2) All artists may see one screening for free. You must sign up for one show before the end of the first read-through or you won't get a comp. We need to know which showing you'll be at so we can sell the right number of tickets and not oversell a particular screening.
3) Please invite people to come to the screening. It's going to be a hell of a lot of fun and if every participant brings ten people we will sell out. This is desirable.

4) You'll be provided with postcards and an email announcement with all the details which you can add a preface to and send to anyone you wish to invite.

II. ACTORS

1) Please bring four headshots & resumes to the first meeting.
2) Know you're going to have a *very* short period of time to memorize your pieces and things can often change on the fly on-set. Roll with the changes and set time aside to memorize the final script as soon as you receive it.
3) Actors must make themselves available for redubbing during postproduction.
4) In extraordinary circumstances, performers may request to withdraw from a project. We ask that the performers not leave a project because they don't like the script, their part, or don't approve of the style, content, humor, or aesthetic of the piece. That's the fun of RIPFest—diving into a place where you might not normally go. We ask the performer only request

to be taken off a piece if he/she finds it morally offensive.

5) Not all parts are created equally. Some have more lines than others. Don't ask for a larger part just because you're feeling yours is smaller than you'd hoped for. Make the part you have memorable by being extraordinary regardless of the number of lines or amount of time on-screen.

III. WRITERS (and composers)

1) There is no need to make all parts equal. If it serves the story, script, or/and film for some actors to have fewer lines and less time on-screen, that's just fine.

2) Write material for the performers assigned to you. Write to their strengths. Really design it like a tailored suit and everyone will look fabulous.

3) Remember you're working in a visual medium. Consider the story, the action, and the juxtaposition of images as much as the dialogue. Let the images tell the story as well as the words. Work with your director on this.

4) Twenty-four hours after the first meeting, have a conversation with your director about your rough draft and get feedback for you to incorporate before the reading the next day.
5) Writers will be responsible for providing single-sided three-hole-punch copies of script for the entire team. Both for the first reading as well as the final draft. How many copies? (No. of performers plus nine—for you, your producer, director, composer, editor, director of photography, a PA/AD, a sound guy, and archives). Please arrive fifteen minutes before your scheduled rehearsals for the first reading to hand out the scripts. *Composers* are responsible for making copies of the music for performers, preferably double-sided and three-hole-punch.
6) Please email an attached MS Word doc of your first draft and final draft as soon as you finish each. Send copies to your entire team, producer, supervising producer, executive producer, and associate art director and artistic director. Email a final

version to reflect all final changes after final editing. *Composers* send final files to David as well.

IV. WRITERS and DIRECTORS

1) All films are limited to ten minutes. Including credits. *No exceptions.*
2) Shorter is better. Always.
3) Cutaways, inserts, exteriors, and establishing shots can be shot anywhere in addition to the two locations you are given. In these sections there can be *no dialogue* or singing.
4) For security reasons we ask that you not shoot those exteriors in or near any subway, tunnel, bridge, police or fire station, or monument including City Hall, Gracie Mansion, etc.
5) We also ask that you not write or include in your film the use of any weapons in an outdoor setting (e.g. no guns [even toy] or knives or even hand fights). Staging using these things in private or/and indoor locations is okay.
6) After the read-through of the first draft, all writers are strongly encouraged to

work closely with the director to create a shootable script. The writer and director should decide immediately after that meeting whom they want to have a postmortem with. It can just be the two of them, or everyone on the team, or anywhere in between.

7) If both the writer and director agree the work is not ready for any kind of public presentation, they may make a request to the producers, who will consider removing the piece from the screening. However, the producers encourage all work, no matter what stage of preparedness, to be presented. We're all about *raw*. And we have an amazingly supportive audience.

V. DIRECTORS

1) If you want to have rehearsals other than those in the basic schedule, work with your producer to decide the time and location for them. RIPFest will not pay for any additional rehearsal space, but some rehearsal space may be available at our offices through the week from 6

p.m. to midnight. (*For movie musicals, Raw Impressions will provide rehearsal space for dance to be arranged with the choreographers and producers. Composers will provide their studios for music rehearsal. If they are not able to do so, contact David or/and your producer immediately.*)

2) You or the producer should email the shooting breakdown/schedule to your entire team the night before the shoot begins.

3) Some PAs may be provided, but you are welcome and encouraged to bring on your own assistant(s). Please give their contact info to your producer via email as soon as you have them confirmed so we can add them to the contact sheet. They must also sign agreements with Raw Impressions. You may give them titles including but not limited to PA, AD, Script Supervisor, etc.

4) You'll have the opportunity to tweak your film after the initial screening. You won't be able to reshoot anything, but you can reedit. Final edit is due two weeks after the

screening. The DVD created will be of the versions of the films that were screened.
5) Get clear on what you *want* to do and what you *need* to do.
6) If there's something you *really* want but don't have, *ask* your producer, they will find a way to tap into the Raw Impressions resources. That includes the relationships and contacts with the 750 people who've participated in Raw Impressions.
7) Simple is better in RIPFest. Always.

VI. DIRECTORS and COMPOSERS
1) Every film will have an original score composed for it. No previously created music copyrighted by anyone may be used unless special arrangements are made with the producers. And that is very unlikely. This includes something as small as a half-second clip. Even if it's from a Bach LP. You need to prove in writing the recording is public domain. Anyone who uses unauthorized copyrighted music will have their film pulled from the screening.

VII. COMPOSERS

1) Bring three copies of a CD with as much music as you like which best represents you and the breadth of your style to give to your director, producer, and editor.

2) You are responsible for creating a deliverable digital audio recording of music for your project. Work it out with your director and editor as to the best format to deliver the score. RIPFest has no budget to hire musicians or recording facilities, though if you can get them for free, you're more than welcome to use them. (Movie musicals is the exception to this last sentence. Arrange all music production with David Rodwin.)

VIII. FINALLY: EVERYONE

1) Thank you for participating. Remember, everyone is donating their time. None of this would work without everyone's amazing commitment and generosity.

2) Don't be afraid of making a dazzling product. This is not just an exercise in collaboration.

3) Don't be afraid to be a leader. Every person has a huge impact on the tone of each collaboration and the eventual product. Even if you never say a word. Even if you're just a PA. You can change the course of your team and project.
4) If you don't know how to do something, that's totally fine. Just tell people so they don't expect you know and they can accommodate that and you don't have to pretend to know something you don't really know.
5) Up your minutes on your cell phone if you can. We're serious.
6) We will provide you with a DVD of all the films. They may not be available for a number of months after the event.

Thanks for reading all of this and have fun.

ACKNOWLEDGMENTS

JOHN

I'd like to start by thanking everyone else that I didn't specifically mention below, so that you don't think my gratitude is somehow lessened because I neglected to mention you by name. You know who you are. Thank you.

This book has been living in my head for over a decade, and I'd like to thank some folks for helping it germinate. First of all, I want to thank my wife, Chrissy, for being the ear, the springboard, the skeptic, the cheerleader, the motivation, and the muse. Thank you.

I'm lucky to have had some great bosses in my career. All of them have taught me something that has worked its way into this book, especially Mark White, John Crandall, Ken Foushee, John Rose, Duwayne Milner, Sheree Thornsberry, Lowell Lindstrom, Maggie Bullington, Ian Eshelman, and Cindy Hembrock.

The original cast of Sketch Development has been an invaluable source of inspiration, whether they know it or not. Calvin Horrell, Cody Frederick, Ryan Jensen, John Quagliata, Heather Davis, John Gruhala, Chris Little—thank you.

The VersionOne gang showed me a better way to think about this subject matter. Lowell, Matt Badgley, Lee Cunningham, Steve Ropa, Maggie: Thank you.

There are a few people who entrusted Sketch Development with their business, and I've learned as much from them as I hope they have from me. Donna Wendel, Jaimee Robles, Alex Basa, Lisa Turnbull, Mark Brooks, Brian Holman, Nate Moore, Kevin Higgins, John Cacavais, Jerret Batson, Leon Green, Kevin Keller, Jeff Villmer, Frank Thoennes, Kim Slate, John Sivill: Thank you.

Scott Kolbe, thanks for describing the book-writing process in a way that made me think I could do it. Frank Hopper, Kacy Wren, Ricki Oldenkamp, Sheila Trask: Thank you for keeping the train moving. Dan Gower, your honest feedback based on an early read made this a better book. I want to thank Sketch consultants, past and present, for contributing their stories. James Nippert, Matt House, Ian Garrison, Steph Weisenbach, thank you.

Tyler Dougherty, Dan Curran, Betsy Irizarry, Brian Warden, Jason Wrubel, Mark Taylor, Mike Venneman, Sung Kang, Joey Frank, David Bogue, Paul Chang, Emily Williams, Keith Hamburg, Matt Kelly, Gary Kupferle, Jeff Barczewski, Keith Compton, Julia Jahnke, Chad Gallant: Thank you for joining me on this journey.

Brian Siedlecki, wherever you are, thank you.

Last but not least, this book would not have been written without Rob Kutner. Literally. Rob, thanks for a great partnership.

ROB

I'd like to start by thanking Nate Roberson at Gotham, who first brought me onto John's radar. This book has been a stimulating, eye-opening, outside-my-comfort-zone odyssey.

Second, I am grateful to Mark McKinney, Dan Powell, David Rodwin, and Dino Stamatopoulos, for their time and insights over some very long and detailed interviews! And also, I thank them for their years/decades of behind-the-scenes (and in Mark's case, in-the-scenes) contribution to advancing the art of sketch comedy and making us all laugh.

Finally, right back atcha, John. This was a collaboration that felt less like a job and more like a year-and-a-half sketch show.

ABOUT THE AUTHORS

John Krewson is the founder and CEO of Sketch Development Services, a custom software development consultancy based in St. Louis, Missouri. Sketch, a two-time member of the Inc. 5000, also provides product management consulting and training inspired by professional sketch comedy production. Since 2015, Sketch and John have helped companies like Mastercard, Nestlé, Centene, and US Bank optimize for agility. As a keynote speaker, John blends his unique entertainment background of acting and producing (including a brief SNL appearance) with his decades of software product leadership experience to guide audiences toward modern ways of working. To learn more about Sketch Development's services, visit *sketchdev.io*.

Rob Kutner is an Emmy-winning writer for late-night shows including *Dennis Miller Live*, *The Daily Show with Jon Stewart*, *The Tonight Show*, and CONAN. He is also a senior lecturer in Comedy Writing at Loyola Marymount University's School of Film and TV.